P9-DFF-716

NORTH DAKOTA
STATE UNIVERSITY
MARY LOU & GEORGE SHOTT
COLLECTION
NDSU
LIBRARIES

American Illuminations

American Illuminations

Urban Lighting, 1800–1920

David E. Nye

The MIT Press
Cambridge, Massachusetts
London, England

This book was set in Bembo by the MIT Press. Printed and bound in the United States of America.

Library of Congress Cataloging-in-Publication Data

Names: Nye, David E., 1946- author.
Title: American illuminations : urban lighting, 1800-1920 / David E. Nye.
Description: Cambridge, MA : The MIT Press, 2018. | Includes bibliographical references and index.
Identifiers: LCCN 2017022787 | ISBN 9780262037419 (hardcover : alk. paper)
Subjects: LCSH: Street lighting--Social aspects--United States--History--19th century. | Street lighting--Social aspects--United States--History--20th century. | City and town life--United States--History--19th century. | City and town life--United States--History--20th century.
Classification: LCC TD195.L52 N94 2018 | DDC 388.3/12--dc23 LC record available at https://lccn.loc.gov/2017022787

10 9 8 7 6 5 4 3 2 1

Overabundance of light produces multiple blindings.

—*Walter Benjamin*

Contents

ACKNOWLEDGMENTS

The research for this volume took me to so many places that it is quite impossible to recall and thank all those who helped me along the way. I have researched the history of electrification in connection with four earlier books, and some materials gathered thirty or even forty years ago have found their way into this volume. Yet the majority of the citations are to sources I encountered only recently, because for the first time I was exploring lighting before 1880 and because a wealth of new sources have become accessible since I researched *Electrifying America*. The first presentation of this work was at the Yale School of Architecture in 2008. As the book proceeded, I spoke at the University of Houston (March 2013), Northwestern University (April 2013), the annual meeting of the Society for the History of Technology (Detroit, November 2014), the University of Pennsylvania (January 2015), the Massachusetts Institute of Technology (September 2015), the University of Virginia (September 2015), the Copenhagen Business School (May 2016), and the University of Hildesheim, Germany (June 2016). The discussions after these lectures helped me to frame the argument and suggested areas for further research.

I thank the MIT Press once again for its high standard of editorial assistance at all levels, and my university for maintaining the sabbatical tradition, which allowed me during the second half of 2015 to spend a month at the University of Virginia, Charlottesville, several weeks at the Smithsonian Museum's Archives Center, and shorter visits at the New York Historical Society and the MIT Library. I benefited from comments and encouragement from many scholars, particularly Paul Israel, Bernie Carlson, Merritt Roe Smith, Miles Orvell, Ronald Johnson, Richard

Hirsh, Martin Melosi, and three peer reviewers. None of these good people is responsible for imperfections that remain. Once again, Helle Bertramsen encouraged my work and helped me to balance writing with the rest of life.

Introduction

Imagine that every night for the last thousand years, a satellite circled the earth taking night photographs. If they were assembled in a ten-minute time-lapse film, one could observe the diffusion of lighting. For the first five minutes there would be occasional spots of light when cities burned down, volcanoes erupted, or forests caught on fire, with long periods of darkness. Fireworks might also be visible occasionally in close-ups of China, and about halfway through the film they would burst forth in Italy and spread to the rest of Europe. In the last third of the seventeenth century, faint streetlights would begin to appear, first in Paris and almost immediately afterward in other cities. But only after 1800 in the last two minutes of the film would the cities be marked by orange-reddish spots from gaslight, first in Birmingham, Manchester, and London, and soon after on the Continent and in North America. This first great energy transition reached cities in much of the world by the middle of the century, notably in the British and French colonies. The number of illuminated places and intensity of the lighting increased for the rest of the century, bursting to a brighter level when electric arc lights were adopted after 1875, and intensifying further as businesses and households adopted incandescent lighting. In the twentieth century, light spread along highways into the countryside.

The film would accurately suggest that cities were the nodal points for two energy transitions, first to gas and then to electricity, both of which reinforced the city's political and economic importance. There were moments of particularly intense illumination, such as world's fair sites that dimmed when they closed down, but otherwise the spread of illumination proceeded with little check to the present day, with the exception of the wartime blackouts in the 1940s. From a satellite, every city can now be located, while much of the surrounding countryside remains obscure. This book is about these two energy transitions as they were experienced in public space. The argument focuses especially on the years from circa 1870 to 1920, when the illuminated urban world was understood to be the pinnacle of civilization's conquest of nature.

The process of illumination was far less uniform than the Olympian perspective of satellite photography might suggest. Not all cities adopted the same lighting systems, and there was considerable variation in energy consumption as Europeans and North Americans tried out different forms of public lighting in the nocturnal landscape. Many different system designs were possible using gas and electrical technologies, and the now-familiar form of US cities had not yet been decided. As late as 1905, only one home in twenty had electricity, and both gas and electricity were first experienced in public spaces that were brighter than most domestic interiors.[1] The night city became more alluring as the hours of commerce expanded and the pleasure-seeking public increased.

How should one understand this process? In a pioneering work, Wolfgang Schivelbusch argued that the "industrialization of light" had "disenchanted night." While his *Disenchanted Night* was full of interesting observations and awakened interest in the cultural history of lighting, the present work takes issue with some of his findings. It is hardly surprising that more than three decades after his groundbreaking book appeared, its conclusions need to be revised or in some cases rejected. Schivelbusch asserted that the centralization of energy systems promoted the centralization of business.[2] More recent work has found that the availability of electrical power often had the opposite effect. Steam engines were expensive and required continual maintenance, imposing high costs and a scale of operation that favored larger enterprises. For example, steam-driven printing presses were too expensive for small print shops, which continued to use muscle power until driven into bankruptcy. Electrical motors helped smaller printers to survive, as these businesses did not need to own and operate an expensive generating system on their premises.[3] The same was true of electrical furnaces and electric lighting for other enterprises. For every enormous illuminated advertisement, there were hundreds of small electric signs that promoted small businesses. The effect of electrical service was frequently to decentralize, whether in the dispersal of people in a household where every room had electric light, the population into suburbs served by electrical subways and streetcars, or electrical tools into small industries that relied on skilled labor. It is true that industry consolidated in this time period, but it does not follow that electricity always promoted this process. The shift in energy regime from steam to electricity made power cheaper, easier to acquire, and more flexible in use. If electricity made possible the assembly line factory, it also enabled many small enterprises to survive and compete successfully, especially where skilled labor and differentiated products were involved.[4] Electricity was an enabling technology, and its effects were complex.

Schivelbusch treats the industrialization of light as a process driven by capitalism with rather uniform effects in Europe and the United States. His book's organization expressed this point of view, as his five chapters treat the lamp, street, nightlife, drawing room, and stage. The organization implied that cultural differences were minor, and streets, shops, homes, and theaters everywhere were much the same. They were not. As I show in *Technology Matters*, summarizing the work of other scholars, technological change is by no means uniform and is shaped by culture.[5] Moreover, a new technology only gradually displaces an old one, in a process examined in chapter 2. There are discernible patterns in energy transitions, whether one examines the shift from gas to electricity, or the shift to renewable energies occurring today. But these transitions are not automatic. They occur at different rates and have different outcomes in different societies.

Gas was not only a new fuel. It replaced sporadic and decentralized lighting with a centralized, uniform system. Nor was electricity simply a replacement for gas. For millennia, burning had always created light, yet electricity was not fire. A gas system transmitted organic energy directly to the consumer, who created some waste products when the gas was burned. An electrical system converted different kinds of energy (wind, falling water, or steam) into a single, inorganic form that passed through a consumer's devices and left nothing behind. The waste was displaced to distant mines, oil wells, power plants, and the like. Gas was organic and direct; electricity was inorganic and indirect.[6]

Schivelbusch not only misunderstood electricity's economic effects, undervalued the extent to which culture shapes technology, and did not conceptualize the shift from gas to electricity as an energy transition but also adopted the dominant metaphor of "disenchantment" to explain the meaning of artificial lighting. This evoked romantic conceptions of what nighttime was once like, as though the world before 1800 had been poetic and enchanted compared to a drab industrial aftermath. Yet study of the public response to gas and electric lighting suggests that lighting itself was often mesmerizing, replacing dull night with enlivening color, and opening up the city to new forms of commerce and pleasure. To see this excitement as disenchantment demands a presentation of the "enchanted" world before artificial lighting, but Schivelbusch's book does not make such a presentation, and his title remains an unsubstantiated metaphor.

Schivelbusch recognized that artificial lighting expanded life into the night, but he missed transatlantic cultural differences. For instance, he treated expositions only briefly, as though they were a uniform phenomenon, and gave the impression that Europe, particularly Paris, set the pace. By 1898, in fact, the lighting at US expositions was more intense, subtle, and artistic, and based on more comprehensive and coordinated planning. Schivelbusch saw the Paris Exposition of 1900 as the epitome of early electrical development.[7] It was not. As explained in chapter 5, technical experts found these Parisian electrical displays to be out of date and incoherent, and the event relied to a considerable degree on gas lighting. Likewise, a reader of Schivelbusch does not learn that until after 1900, London's streets were mainly lighted with gas, in dark contrast to intensely electrified New York. Nor did Schivelbusch recognize that in 1903, the per capita consumption of electricity in Boston and Chicago exceeded that in Paris or Berlin by 400 percent. Such an enormous technical difference expresses a cultural contrast, for the functional and symbolic aspects of illumination were always intertwined.

This book asks why Americans developed such intense urban lighting and how they used it to shape their urban culture, as compared to European developments. The first chapter examines the Renaissance tradition of illuminations and civic celebrations, which persisted in Europe and spread to the United States, where it merged with a vigorous parade tradition. Chapter 2 reviews the two energy transitions—to gas and then to electricity. Chapter 3 explores why, despite having access to the same technologies, Britain and the United States developed different public lighting systems. Chapter 4 looks at tower lighting, a US system briefly widespread in the Middle West, South, and West that expressed a different aesthetic and value system than the now-familiar rows of streetlights that line US streets. Chapter 5 examines spectacular lighting at expositions from Paris in 1881 to Buffalo in 1901 as well as Saint Louis's annual Veiled Prophet celebrations and other regional events. These urban spectacles increased interest in lavish, permanent lighting installations. Chapter 6 discusses the commercialization of public space using both gas and electric lighting, culminating in giant advertising signs, scintillating downtowns, and the dramatic lighting of skyscrapers, bridges, and public monuments. By 1900, this nocturnal landscape was a hallmark of popular culture. Yet many in Europe and some in the United States considered this cityscape garish as well as visually incoherent, and chapter 7 explores the City Beautiful movement's efforts to create a more harmonious aesthetic at events such as the Hudson-Fulton Exposition of 1909 and a series of expositions that culminated in 1915 in San Francisco. Chapter 8 then turns to how spectacular lighting became a part of the US political system, including parades, presidential inaugurations, and the lighting of national symbols like the Statue of Liberty. The final chapter concludes the argument that the forms and uses of public lighting were by no means inevitable. Only after considerable experimentation was spectacular lighting made a seemingly natural part of streets, skyscrapers, landmarks, and events. However inevitable they may seem today, they are social constructions that express political and social values.

These chapters concern a few European cities, notably London and Paris, and the fifteen largest US cities in 1900: New York (3.4 million), Chicago (1.7 million), Philadelphia (1.3 million), Saint Louis (575,000), Boston (561,000), Baltimore (509,000), Cleveland (382,000), Buffalo (352,000), San Francisco (343,000), Cincinnati (326,000), Pittsburgh (321,000), New Orleans (287,000), Detroit (286,000), Milwaukee (285,000), and Washington (279,000).[8] At the time, the transformation of night space in these cities was thought to have specific, desirable consequences. It was widely believed that illumination was a check on crime, and that it expanded the public sphere, not least for women.[9] More generally, public lighting was regularly discussed in terms of social uplift, especially during the Progressive era, notably at world's fairs and among supporters of the City Beautiful movement. Light seemed to measure a city's progress. Command of energy was understood to be essential to technical advances, prosperity, and a higher level of culture. When H. G. Wells visited New York in 1906 he thought, "New York is lavish of light, it is lavish of everything, it is full of the sense of spending from an inexhaustible supply."[10] The apparently perpetual supply of energy seemed to guarantee a cornucopia of goods as well as round-the-clock stimulation and excitement.

Positive views of electrification were qualified both in the nineteenth century and after. During the Progressive era, electricity increasingly came from coal-fired power plants that polluted the air and threated human health. The plants reduced sunlight while exposing the lungs to noxious gases and soot particles.[11] Fears of air pollution were not strong enough in either Europe or the United States, however, to prevent extensive coal burning to produce gas and electric light.

To a degree, social elites employed electrical displays to project their social status and justify their power. Electric lighting was more than an overt means of social control that made the city more visible to the police. In addition, from the Renaissance onward, the court, aristocracy, and urban elites used electrical displays to transform the appearance of the city as well as to excite awe and admiration. But one must not suppose that hegemonic intentions were always realized. Had local elites in the United States been more unified, they might have imposed a uniform pattern of illumination with a coherent aesthetic such as that at a world's fair. In practice, though, electrification was often a matter of consumer choice. Through imaginative lighting, small businesses created distinct identities. Neighborhoods used lighting displays to express a local identity, such as New York's Chinatown or Little Italy. Electricity became a tool of self-expression that to a considerable degree resisted the efforts of reformers to make over cities to resemble the great expositions. US electrification was frequently less hegemonic than individualistic, expressing a mosaic of social worlds, and resulted in lively landscapes of consumption such as Coney Island and Times Square.

The appearance of the night landscape was not always foreseen or controlled. The large number of actors, public and private, large and small, produced unexpected juxtapositions and transformations, notably in city centers where businesses vied for attention, and large electric signs were continually being erected and replaced. The effect of this landscape on the average citizen could be defamiliarization. By night, the city was radically altered. Many in the City Beautiful movement bemoaned what they saw as the incoherence and visual cacophony of an increasingly commercialized public space.[12] Others, such as Ezra Pound, embraced New York at night as an expression of modernism. "No urban nights are like the nights there; I have looked down across the city from high windows. It is then the great buildings lose their reality and take on their magical powers."[13]

Michael Foucault argued that modern civilization created a hybrid form of social space, which he called "heterotopia," which "is capable of juxtaposing in a single real place several spaces, several sites that are in themselves incompatible."[14] The great expositions enacted such a heterotopian transformation each evening, as visitors saw elaborate lighting effects alter the entire grounds. A cadre of engineers became specialists in creating such transmutations, staged as performances just after darkness had fallen. Foucault specifies that "heterotopias are most often linked to slices in time," and begin "to function at full capacity when men arrive at a sort of absolute break with their traditional time."[15] Few things could be more traditional than nightfall, which had always divided human experience into two quite-distinct periods. To break through the darkness with massive lighting displays erased this demarcation and declared human independence from the rhythms of nature. The illuminated city, with its bright boulevards, skyscrapers, and spectacular electric signs, seemed to exemplify progress, representing the triumphant light of civilization. Like the world's fairs, illuminated cities were widely understood to be dynamic utopian landscapes, while an unelectrified city was backward.

Americans long celebrated electrification as a turning point in historical experience. In 1939, Consolidated Edison's exhibit at the New York world's fair depicted electrification as a break in space and time.[16] Visitors walked along a full-sized cobblestone street that depicted 1892, where they saw small shops, an ice wagon, a horse-drawn streetcar, and gas streetlights. They glimpsed housewives washing clothes by hand or cooling themselves with handheld fans. After this "Street of Yesterday," they entered the brilliantly illuminated "Avenue of Tomorrow" with smooth asphalt streets, sleek automobiles, and skyscrapers with well-lighted plate glass windows. The two landscapes told the public that electrification had broken the grip of darkness and ushered in the modern world. Yet the transition from gas to electricity had not been as sudden as this exposition pavilion suggested, for it had required half a century.

Moreover, illuminated modernity was fraught with contradictory implications. As T. J. Jackson Lears demonstrated in his magisterial *No Place of Grace*, many Americans were uneasy with the cultural effects of the second industrial revolution. Particularly the educated and upper class began to idealize the preindustrial past, cultivate arts and crafts, celebrate medieval cathedrals, retreat to pastoral settings, or sojourn in Europe, where the rush of progress seemed less pressing. But "by exalting 'authentic' experience as an end in itself, anti-modern impulses reinforced the shift from a Protestant ethos of salvation through self-denial to a therapeutic ideal of self-fulfillment in this world."[17]

Electrification expressed contradictory possibilities. On the one hand, the United States was a commercial civilization, where businesspeople, joined by engineers and progressive reformers, sought to increase society's rationality and productivity. Mastery of electricity assisted this rationalization, whether by controlling the flow of street traffic with semaphores, lengthening the workday with new lighting systems, speeding delivery of messages through the telegraph and telephone, or packing more experiences and more production into each day. On the other hand, electricity increased the allure of new pleasures that detracted from the work ethic and self-control at amusement parks, vaudeville shows, dance halls, pleasure gardens, rooftop restaurants, the cinema, and much else. The metaphor of disenchantment is inadequate to describe these contradictions. The electrified city was the site of both rationalization and the intensification of pleasures. It was both a productive dynamo and the Great White Way. Illuminations were woven into urban cultural life, and found expression in commerce, architecture, landscapes, spectacles, patriotic events, tourism, entertainment, and politics. Taken together, the many cultural uses of lighting created a heterotopian city that was alluring yet undefinable, more vivid and yet less concrete, perpetually bright and yet transient in its details. The city had become both a vast mechanism that hummed through the night and an undying fireworks.

1

Illuminations

During the Renaissance, royal courts and cities held illuminations on special occasions. The purpose was not utilitarian but rather ceremonial, to glorify a monarch, mark an anniversary, or celebrate an event. Illuminations later were partially submerged in the sea of public lighting, but in the early modern period they were powerful precisely because the city was normally dark at night. Elaborate lighting was a mark of distinction reserved for elites, and European courts developed a tradition of fetes and illuminations, particularly in Italian city-states, notably Florence. Vannoccio Biringuccio (1480–1539) in the tenth section of *La Pyrotechnie, ou, Art du feu* provided much of the essential information. He presented fireworks in mythological terms, as an extension of ancient Greek practices, the explosive meeting of fire and air, or Vulcan and Venus.[1] His often-translated work helped spread fireworks and illuminations to the rest of Europe. By the baroque period, the association of illuminations and fireworks with ancient Greek festivals was well established. In a technical manual written in the midst of the French Revolution, Claude F. Ruggieri presented fireworks as an extension of ancient custom, noting that "the Greeks conserved fire, as a divine essence, in their sacred places."[2] Ruggieri explained how to make roman candles, flaming serpents, pinwheels, rockets, and "buckets of fire." There was also a section on how to adapt fireworks to theatrical productions.

Beginning in 1471, the papacy sponsored a spectacular fireworks display called the Girandola at the Castel Sant'Angelo in Rome, the papal fortress originally constructed as the mausoleum of the emperor Hadrian. These fireworks would be staged every year at Easter, on the eve of the Feast of Saints Peter and Paul (June 28) and whenever a new pope was elected (see figure 1.1). Architects took charge, creating "elaborate temporary edifices for fireworks" that included "fantastic imaginary structures, full-scale temples in three dimensions, constructed with wood and iron frames, and hung with *trompe l'oeil* painted cloth and papier-mâché and stucco decorations prepared by a small army of carpenters, turners, painters and sculptors." These were reputedly "the most elaborate fireworks displays of the eighteenth century."[3] Johann Wolfgang von Goethe wrote of seeing "La Girandola" on June 30, 1787. Fireworks, torches, and lamps brilliantly illuminated the Castel Sant'Angelo and the prominent buildings of the Vatican. "To see the colonnade, the church, and, above all, the dome first outlined in fire, and after an hour becoming one glowing mass is a unique and glorious experience." Goethe thought the fireworks were beautiful, but was mesmerized more by the "illuminations of the church," which had been turned "into mere scaffolding for the lights."[4] He witnessed how dramatic lighting could transform a site and lend it an air of enchantment. Yet the more frequently such events were held, the weaker their impact, unless they increased the size and brilliance of the displays or included new special effects. The lighting that captivated Goethe would seem tame to later generations.

Peter the Great became a patron of pyrotechnic displays and enthusiastically took part in firing off rockets before his assembled court.[5] Fireworks were also popular at the French and English courts, and by the end of the eighteenth century they were the expected finale to any important public event. "Seventeenth and eighteenth century court society" in France "relied on great quantities of artificial light to confer distinction upon their nocturnal existence." But "the light mobilized by the court, either at Versailles or in urban aristocratic *hotels*, was of a rare and exceptional sort, small exclusive pockets of luminosity" that underlined the status of those within its circle, while eclipsing the rest of society.[6] One of the most spectacular displays of the eighteenth century was witnessed by a reported six hundred thousand spectators in Paris on the occasion of the marriage of Louis XV's sister to the heir to the Spanish throne.[7] Another was held in London in 1763 to celebrate the British victory over France in the Seven Year's War (see figure 1.2).

Figure 1.1 Francesco Piranesi, La Girandola, 1758.
Source: New York Metropolitan Museum of Art

1.2 Fireworks in Green Park, London, 1763
Source: Danish National Library, Copenhagen

Well after the adoption of illuminations, in the second half of the seventeenth century the major European cities installed oil streetlights, starting in Paris in 1667. These streetlights were soon taken up in Britain, the Netherlands, Germany, and central Europe.[8] The cities became more navigable at night, and a new public life began to emerge. Where nighttime had been focused on private life in the home, cities started to develop new institutions, notably coffeehouses, theaters, and restaurants, but also societies, associations, and evening lecture series. As people increasingly met at night in public, what Jürgen Habermas later termed "the public sphere" emerged.[9] Regular public lighting expanded the geographic extent and temporal accessibility of this sphere. Lighting was necessary for the public sphere, but not sufficient for it to emerge. A city with salons, associations, and cafés also required widespread literacy, some freedom of expression, and a middle class.

In the eighteenth century, street lighting was still modest. Paris in 1763 had almost six hundred thousand people, yet only sixty-five hundred public lanterns.[10] London was slightly better served, but still dark in many places. The American colonies were darker. In 1697, New York's city magistrates noted "the great Inconveniency that attends the Citty, being a Trading Place for want of Lights." They attempted to remedy the problem by ordering every homeowner to hang a light on a pole from a second-floor window. When the citizens objected this would be costly, the order was modified to one light for every seventh house, to be provided during "the dark time of the moon."[11] American towns remained dark until Benjamin Franklin organized lighting with oil lamps in Philadelphia in 1751.

Proponents of street lighting often argued or implied that before artificial illumination, the city went to sleep after nightfall. Somnolent streets, deserted except for the night watchman, were a staple in this narrative, in which economic and social life came to a standstill until dawn. Historians have discredited this story, for people remained active during the evening hours.[12] Ordinary people divided their tasks into those that had to be done by day, and tasks that demanded less light and could be performed at night. Some trades, such as baking or garbage collection, were primarily night activities. Circumstances dictated many tasks. A child or foal might be born at 2:00 a.m. A patient might need medical attention in the wee hours. Many amusements took place mostly at night, such as workers drinking in a tavern, farmers holding a dance, or high society attending a ball. This in turn meant that bartenders, servants, cooks, carriage drivers, and many others had to work at night. Streetwalkers also plied their trade under cover of darkness and were called "ladies of the night." Clearly, their clients were also abroad. Farmers rose before dawn and carted their produce to city markets, where customers began to arrive as day was breaking. The night city was not entirely asleep before gas and electric lighting awakened it to modernity. Many were out and about, and some activities were confined to night. As A. Roger Ekirch has noted, "Night time commonly blurred the boundaries between labor and sociability" when people might spin, knit, thresh, husk corn, or go fishing.[13] At night people also found time to play cards, sing, and tell stories, frequently accompanied by a sociable drink. Darkness partially leveled society and loosened social controls, and those who were subservient by day enjoyed greater freedom during the night. Secret societies, protest groups, homosexuals, and persecuted religious sects commonly met under cover of darkness.[14]

The first public lighting in Europe was intended to aid police in preventing crime, light the way for pedestrians, and enhance the dignity of central locations. Businesses might stay open a little longer, but this was not at first presented as its chief justification. Yet public activity did increase as illumination intensified. In this "nocturnalization of urban daily life," the theaters, clubs, and restaurants opened later. Plays in Restoration London began at 3:00 p.m. and ended as it was growing dark; by the end of the eighteenth century, they started at 6:30 p.m.[15] Simultaneously, it became difficult to enforce the traditional curfew in London and other large cities. When gaslight was introduced a century later, nocturnalization expanded into later hours.[16]

Oil lamps provided faint light compared to the spectacular celebrations that accompanied coronations, weddings, military victories, and commemorations. Louis XIV consistently used fireworks and illuminations to entertain the people, awe visitors, and confirm his right to rule.[17] At Marie Antoinette's marriage to the king of France, the fireworks lasted for half an hour, "and included hundreds of rockets and thousands of explosions, along with 2000 Roman candles, turning stars, and jets of fire."[18] Many illuminations were not staged by the court but rather combined private displays into a civic ceremony. Governors or mayors typically called for these spectacles, yet their success required citizens to take an active part. Government buildings and palaces were decked with lanterns, often accompanied by fireworks, but homeowners, organizations, and businesses contributed further lighting effects. These events were all the more memorable because of their contrast to the normally dark city. As Craig Koslofsky explains, "The illumination placed multiple lights in the windows of a single building or across an entire city, a massive yet precise display of loyalty and obedience to the ruler."[19] As illuminations became widespread in the seventeenth century, cities as well as courts began to organize them. In 1699, the citizens of Madrid proposed illuminations to celebrate the return of their king from a long period at his palace in Escorial. Two years later there were illuminations to celebrate his visit to France, and in 1711, on his behalf Madrid staged illuminations and "machines of fireworks of a surprising beauty."[20] By this time illuminations were so familiar that British newspapers scarcely described them, and a typical announcement read, "To all the Nobility and Gentry, in Honour of the First Christmas of her Majesty's Reign," there will be a "Magnificent Entertainment of Musick, with Illuminations after the true Italian manner."[21] In Naples, a saint's day in 1703 was celebrated with "the usual Rejoycings and Illuminations," which included discharge of "all the artillery of the Castles." By the early eighteenth century, German specialists in *Zeremonialwissenschaft* (ceremonial studies) published books on the relationships between authority and such spectacles.[22]

A century later, Napoléon celebrated his marriage with "les illuminations les plus remarquables."[23] Not long after, Britain celebrated its victory over Napoléon with illuminations. These included a wooden pagoda covered with gas jets beside the lake in London's St. James's Park that appeared to be a tower of fire reflected in the water. When it caught fire and burned to the ground, the public applauded, thinking it was a part of the performance. By the mid-nineteenth century, civic celebrations began to include both gas and electric light. As early as 1844, Parisians enjoyed the spectacle of an arc lamp in the Place de Concorde, and Londoners witnessed something similar in 1849.[24] Combinations of gas and electric illumination became common after 1875. European ecclesiastical, political, and royal traditions of illumination did not reject one form of light for another but rather exploited all the possibilities. As Dietrich Neumann observed, "The nineteenth century is characterized by a fruitful coexistence of different lighting technologies, which were sometimes used in conscious contrast."[25]

Yet in the mid-nineteenth century, effective illuminations still could be staged without either gas or electricity. When Mark Twain visited Venice in the late 1860s, he witnessed a fete that relied entirely on lanterns and voluntary private fireworks. The whole city seemed to be boating on the lagoon, and

> in one vast space—say a third of a mile wide and two miles long—were collected two thousand gondolas, and every one of them had from two to ten, twenty and even thirty colored lanterns suspended about it, and from four to a dozen occupants.

> Just as far as the eye could reach, these painted lights were massed together—like a vast garden of many-colored flowers, except that these blossoms were never still; they were ceaselessly gliding in and out, and mingling together, and seducing you into bewildering attempts to follow their mazy evolutions. Here and there a strong red, green, or blue glare from a rocket that was struggling to get away, splendidly illuminated all the boats around it. Every gondola that swam by us, with its crescents and pyramids and circles of colored lamps hung aloft, and lighting up the faces of the young and the sweet-scented and lovely below, was a picture; and the reflections of those lights, so long, so slender, so numberless, so many-colored and so distorted and wrinkled by the waves, was a picture likewise, and one that was enchantingly beautiful.[26]

Neither gas nor electricity was needed to stage this event. The city decreed the spectacle, but the arrangements were spontaneous and individual. Lamps, candles, and fireworks were quite sufficient.

Nineteenth-century Americans on the grand tour through Europe witnessed festivals and special events whose beauty and grandeur relied on a mastery of lighting. This pageantry became part of their definition of culture and was incorporated into their image of the European past. When Twain wrote *The Prince and the Pauper*, he imagined one such event, in which a royal barge "took its stately way down the Thames through the wilderness of illuminated boats" and "the city lay in a soft luminous glow from its countless invisible bonfires; above it rose many a slender spire into the sky, encrusted with sparkling lights [that] seemed like jeweled lances thrust aloft."[27] The science journalist Walter Kaempffert was also aware that "one of the earliest forms of ornamental street-lighting was of a spectacular character for special occasions."[28]

The demand for fireworks increased during the eighteenth century, and fireworks became a commodity available well beyond the court. Using them along with Chinese lanterns, oil lights, and candles, institutions, noble families, and wealthy individuals participated in illuminations, thus calling attention to themselves. As fireworks spread throughout Europe, an unplanned process of democratization ensued.[29] Illuminations ceased to be a royal gift to the populace. By the late eighteenth century, fireworks and theatrical lighting were used in pleasure gardens open to the paying public in Paris and London.[30] Spectacular lighting remained a featured event at the Royal Vauxhall Gardens in the 1840s (see figure 1.3). British illuminations were often organized by cities at the request of the mayor. For example, when Londoners received word of Lord Admiral Nelson's victory over the French fleet, they decked their buildings with lighting. The Admiralty placed a double row of plain lamps along the front of its building, surmounted by flambeaux. There was also "a triumphal arch," and above it "G.R." and a brilliant star. Towering over this display was an anchor of considerable magnitude, supported on each side by a pillar. "The Imperial Crown of Great Britain surmounted the whole."[31] Covent Garden Theater erected a large "letter N. surmounted by the British Crown, encircled by branches of laurel, composed of variegated lamps." The Drury Lane Theater had an equally lavish display. *Oakley's Magazine* on Old Bond Street lined the front of its building with "brilliant lamps" and there were further illuminations in the windows. There were elegant displays on Pall Mall, St. James Street, Mortimer Street, and elsewhere. It was impossible to walk through London and overlook the tributes to Lord Nelson. A decade later, an even more elaborate celebration occurred after Wellington's victory at Waterloo. It encompassed both East and West London, but included little gaslight, as service was not yet available in much of the city.[32] The gas lines reached the Covent Garden area, for instance, only in 1815.[33]

A British fireworks handbook from 1824 enthused over how fireworks moved "with velocity through the air, throwing out innumerable sparks or blazing balls, which fly off into the infinity of space; others, suddenly exploding, scatter abroad luminous fragments of fire, which are trajected with the most speedy trepidation; and again, others are revolving on a quiescent centre, and by their revolutions produce the most beautiful circles of fire, which seem to vie with each other in their emanations of splendour and light."[34] Fireworks were also popular in the United States, where they were combined with lighting, displays of national symbols, and largely spontaneous public participation. A city government could offer suggestions, but it could not control the details

1.3 Poster, Royal Vauxhall Gardens, 1840s
Source: Library of Congress, Washington, DC

of such an event. When President James Monroe came to New England in 1817, for example, private citizens in Boston put up many lights and mounted firework displays to celebrate, and in Portland, Maine, "the illumination was very general and splendid."[35] Likewise, when the Erie Canal was completed on November 4, 1825, many homes and businesses participated in the illumination, which culminated in a magnificent fireworks display held at New York's City Hall. It required 1,542 candles, 454 lamps, and 310 variegated lamps, and included "Chinese and Diamond fires" on the roof, seen against the background of 1,500 "brilliant stars, intersecting each other in fanciful directions." The spectacle concluded with 374 rockets shot into the sky. "The general bursts of simultaneous applause from a great concourse of citizens afford the best panegyric on the decided superiority of these fireworks both as to extraordinary grandeur and brilliant display" (see figure 1.4).[36]

Illuminations might also celebrate a peace treaty or honor a famous visitor.[37] When Queen Victoria made her first visit to Brighton in 1837, for instance, the city's illumination "called forth all the energy and invention of the inhabitants. Never was there so beautiful and brilliant a display of devices in lamps, gas, transparencies etc., seen in this town … and the stars hid their diminished fires."[38] Every bootmaker, hairdresser, stationer, tailor, doctor, saddler, and manufacturer erected gas fixtures or arranged "variegated lamps" to resemble crowns, wreathes, stars, or the letters V. R. The many transparencies showed such subjects as the "Queen, in a car, drawn by three lions, Britannia at their side." Another, considered particularly splendid, was a 250-square-foot transparency, "representing two Corinthian fluted columns, with bands of flowers twining round the shafts, the rose, shamrock, and thistle on the pedestals, and the capitals surmounted with wreaths of laurel, the initials V. R. in emblazoned stars, and surrounded with the national colours; the centrepiece, a man of war with the Union Jack, waving over it; motto, 'Welcome, Victoria.'"[39]

Illuminations were frequent, and the British were adept at staging them on short notice. Some visitors from Hamburg who happened to be in Leeds in 1855 were astonished "that so handsome an illumination could have been called forth with so short a time for preparation" in response to a request from the chief magistrate. This being a celebration of peace at the end of the Crimean War, they were also impressed by "the

1.4 Erie Canal Celebration, 1825
Source: New York Public Library

unsurpassable order and general stillness that reigned in all the streets, although crowded by thousands."[40]

Yet illuminations could serve partisan ends or provide opportunities for the lower classes to vent their feelings. In 1706, one citizen of London asked the mayor to discontinue illuminations because "it gives ye Rude Rabble Liberty to Doe what they list. … [T]hey Breake windows with stones, fire Gunes," and assembled in unruly crowds. Such complaints were voiced during the rest of the century.[41] During the reign of George III there were many illuminations, including one grand event at the Bank of England in 1789 to celebrate his recovery from illness. On some occasions, the illuminations were demanded as an expression of political allegiance. At times, however, crowds destroyed the transparencies and smashed windows in protest.[42] The potential for disorder was well established before 1820 when supporters of Queen Caroline distributed handbills and posters calling for illuminations on her behalf. The working classes preferred the queen to her husband, King George IV. The couple had been estranged for years, and he had introduced a bill into the House of Lords that would turn their separation into divorce. Many people considered this disgraceful, and the question agitated the nation. Those who supported the queen called for illuminations and attacked the homes of those whose windows remained dark.[43] The Edinburgh authorities proclaimed that no illumination should be held because of "the serious riots which have taken place in London, Glasgow, and other places."[44] Nevertheless, the illuminations began at six in the evening in the Old Town. A large crowd assembled as darkness fell, and by 9:00 p.m. they roamed the streets, looking for homes or establishments that did not prominently display candles or lanterns in their windows. In fashionable New Town "they found several of the streets in a state of darkness" and "the windows which were not illuminated were assailed by showers of stones, and the crashing of the glass was heard in all quarters."[45] The police called on "the cavalry from the Riding School [to] disperse the mob." Antipathy to George IV was not limited to the cities. In the town of Ely, for example, few were enthusiastic about his coronation. Ely's illuminations were scant, and although "three barrels of beer were given to the populace," they, "unmindful of this liberal allowance, began immediately to cheer for the Queen."[46]

A decade later, the cause of reform was supported by illuminations. On May 21, 1831, most of Dublin was brilliantly lighted, except for the homes of Tories. The Archbishop's Palace was dark, for instance, and a "volley of stones from some idle boys soon ventilated the house pretty liberally."[47] In Edinburgh, the Reform Bill of 1831 brought out a partisan crowd for an illumination, and as a Tory newspaper put it, "Several corps of young blackguards, who had previously filled their pockets with stones ... paraded the principal streets, chiefly in the New Town, dealing destruction as they went, to all windows not lighted up, and to many which were lighted, but which did not display a number of candles sufficient to satisfy the mobocracy."[48] London was more peaceful. The lord mayor called for an illumination to support the cause of reform and celebrate the dissolution of the unpopular Parliament. These illuminations "were very general, and upon an exceedingly brilliant and splendid scale," including many shops on Oxford Street, Regent Street, the Strand, Covent Garden, and Fleet Street as well as in the major theaters. The "streets were crowded to excess, and among the spectators were many persons of highly respectable appearance." The crowd was not "anxious to create any disturbance or commit any outrage, excepting in those cases where persons obstinately refused to exhibit any illuminations."[49] Many displays focused on the king, and few specifically mentioned the Reform Bill, but these meant the same thing to the crowd. As one Oxford Street shoemaker's display put it, "The King, the People, and the best of the Nobility are united for Reform."[50] London was often unruly during illuminations, however, and a letter in the *Times* complained of "the illumination of private houses as a nuisance, which no one would incur if it were not for the fear of a stone through his drawing-room window."[51]

Illuminations in Ireland could pit Catholic against Protestant. In 1849, during the famine, the lord mayor held an illumination in Dublin, spending £8,000 on candles alone. Nor was this all. He requested illumination of every household at its own expense. One newspaper remonstrated that "amid famine, disease, and death, the Irish capital is to be lighted up to simulate prosperity and joy."[52] It seemed as sensible as "illuminating a graveyard." In 1863, another requested illumination turned the students of the Catholic university against the authorities. Reportedly, "the whole body of the students revolted," saying that any lighting up of the university was unacceptable. They cut the gas pipes and ruined the "illuminating material," not once but twice, preventing the college from participating.[53] Anyone who illuminated a house to celebrate the British sovereign risked broken windows.

While partisan illuminations and attacks on property might have been widely adopted in US political culture, such was seldom the case. The United States adopted other aspects of the illumination tradition, including the use of transparencies, the lighting of public buildings, and fireworks. The early illuminations were little more than placing lamps and candles in windows in order to brighten the street. By the time of the American Revolution, "transparencies" had begun to appear. These "were created by a variety of local people—artists like Charles Wilson Peale in Philadelphia, sign painters, scene painters employed at the theaters," and many amateurs.[54] They painted on light fabric, thinner than canvas, but similarly stretched on frames. These paintings of royalty, and later of revolutionary heroes or patriotic emblems, were placed in windows along a parade route or in a public square, and lighted from behind using candles or lamps. Although faint by modern standards, they stood out in an otherwise-dark street. Transparencies might seem precursors of electrical advertising signs, but they were ephemeral and ceremonial, not commercial.

These practices were modified by the gradual adoption of gas street lighting. Some observers worried that this improvement undermined the tradition of illuminations. As gaslight brightened the night city, it reduced the contrast between its normal and festive appearance. A German architect summarized what was possible using gas: "What a glorious discovery is the gaslight! How its brilliance enhances our festivities, not to mention its enormous importance to everyday life. … [W]e pierce the pipes with innumerable small openings, and all sorts of stars, firewheels, pyramids, escutcheons, inscriptions, and so on seem to float before the walls of our houses, as if supported by invisible hands." Yet there was a trade-off involved when adopting gas for illuminations. The same writer observed that the public became more passive: "Who can deny that this innovation has detracted from the popular custom of illuminating houses as a sign the occupants participate in the public joy?" Before the adoption of gas for spectacles, "oil lamps were placed on the cornice ledges and window sills, thereby lending a radiant prominence to familiar masses and individual parts of the houses." But with gaslight, "our eyes are blinded by the blaze of those apparitions of fire and the facades behind are rendered invisible." On the whole, he concluded, "the art of lighting has suffered a rude setback by these improvements."[55]

Gas lighting made illuminations more brilliant and yet paradoxically less spectacular, because night cities became routinely visible. Gaslight metamorphosed "city space as a whole, reconfiguring nocturnal life into a new artificial universe."[56] Gas added a special aura to the city, making it more visible without obliterating the night. It brightened urban space, transforming it into a stage set, as "the brightly lit boulevards were appropriating the old aristocratic privilege of an overabundance of light."[57] The advent of electric lighting, because it was brighter than gas, further increased this theatricalization. Buildings became scaffolding to hold light bulbs, and like the gas jets before them, they could obscure a building more than illuminate it. Some architects objected, but the public enjoyed the novelty of seeing familiar surroundings transformed.

If illuminations appealed to a broad spectrum of people, building a shared sense of civic pride, they could become the property of the elites who could afford them. Yet many US events remained broadly democratic. During the first half of the nineteenth century, the parades included not only political and military leaders but also featured large numbers of artisans, police officers, and firefighters. By the 1840s, a New York parade's format included politicians "in open carriages, soldiers, more than fifty companies of firemen, the butchers, and the printers." Other groups usually were present as well, including temperance societies, university students, and representatives of civic institutions.[58] Troops became more prominent in parades after the Civil War. Parades were particularly numerous in New York City, and usually took place during the day. There were occasional torchlight processions, but civic ceremonies typically began with the ringing of church bells in the morning, a parade in the middle of the day, and an evening dinner or "cold collation" for the select. A parade might celebrate the opening of a new water supply, dedicate a monument, observe an anniversary, or receive a famous visitor, but the format was much the same. When festivities continued into the evening, they featured bonfires, fireworks, and illuminations.[59]

By the 1820s, illuminations were an established part of US political and social life. In response to the visit of the president or a famous guest, such as the Marquis de Lafayette, Americans had rituals ready for the occasion. Likewise, July 4 became an annual ritual—one that might also mark the completion of a canal, railroad, or bridge. As John Seelye explains, Americans developed "a tradition of oligarchical orchestration dating back to the time of the Revolution and epitomized by that uniquely American secular festivity, the Fourth of July." By the early nineteenth century, "such events were associated with the kind of art works requiring planning and production—illuminations, panoramas, allegorical paintings, floats, triumphal arches, fireworks."[60] Such celebrations were held when the Erie Canal opened in 1825, the Baltimore and Ohio Railroad was begun, the first bridge was completed over the Mississippi, and many other occasions.[61] On the night of August 17, 1858, New York "blazed with illuminations, from tallow dips, gas jets, rockets, and all manner of pyrotechnic displays, all in honor of the [new telegraph] Cable that connects ours and the Mother Continent." It was like a "compacted Fourth of July, or a dozen of them rolled into one."[62] Illuminations also were held to mark the end of the Civil War, with Union Square particularly well lighted. Elaborate "pieces" represented the firing on Fort Sumter, a naval bombardment, and "an eagle and a shield, with the national colors festooned around." At the center was the motto "One Country, one People, one Destiny," while on the side were the names of Union Army generals. The largest display piece represented the naval battle between the *Monitor* and the *Merrimack*.[63]

Illuminations were prominent in political campaigns too. After Abraham Lincoln was nominated for president, the Republicans staged "Wide Awake" torchlight parades and fireworks. Charles Francis Adams later recalled that the "campaign of 1860 was essentially a midnight demonstration—it was the 'Wide-awake' canvas of rockets, illuminations and torch-light processions. Every night was marked by its tumult, shouting, marching and counter-marching, the reverberation of explosives and the rush of rockets and Roman candles."[64] The National Museum of American History preserves a "three-sided transparency," twenty-seven inches high, that was made by stretching fabric over a wooden frame, and then painting it with campaign slogans and symbols—in this case, a portrait of

"Old Abe—Prince of Rails." It was "illuminated from inside by a small oil lamp and carried" aloft on a pole in campaign parades.[65] A phalanx of marchers carrying them had a dramatic effect. Parades with illuminations were also a popular part of the centennial celebrations on July 4, 1876 (see figure 1.5). The political parties embraced illuminated parades in the election that year, when Boston's Republicans and Democrats marched in competing parades bearing torches by the thousands.[66]

Parades were by definition ephemeral, but gas lighting made it possible to illuminate prominent buildings not just for special occasions but permanently as well. In Washington, DC, the Capitol was lighted by gas jets, which Walt Whitman saw in 1865, around the time of Lincoln's second inauguration. "Tonight I have been wandering awhile in the Capitol, which is all lit up. The illuminated rotunda looks fine. I like to stand aside and look a long, long while, up at the dome; it comforts me somehow." He viewed the White House on another evening walk: "To-night took a long look at the President's house. The white portico—the palace-like, tall, round columns, spotless as snow—the walls also—the tender and soft moonlight, flooding the pale marble, and making peculiar faint languishing shades, not shadows—everywhere a soft transparent hazy, thin, blue moon-lace, hanging in the air—the brilliant and extra-plentiful clusters of gas, on and around the facade, columns, portico, &c.—everything so white, so marbly pure and dazzling, yet soft."[67] Lighting made public buildings stand out against the night, augmenting their symbolic role.

After the Civil War, illuminations became even more common. Harvard College began to hold them annually in 1874, for example, and at many "seaside watering places they [were] now generally made the closing feature of the season."[68] Boston held a grand illumination in 1875 to celebrate the centennial of the Battle of Bunker Hill. During a perfect June day, a ten-mile procession wended its way through the city, and at night the public's "eyes were feasted with illuminations and pyrotechnics" that featured "the illumination of the State House," which was "produced by gas and calcium lights. The gas was thrown out in jets from pipes perforated at even distances along their length, and these pipes were extended along the upper cornices of the main structure." Above this display, "brilliant bars of light were thrown athwart the atmosphere [with] calcium lights, using both white and colored lights."

1.5 Illumination of Madison Square, New York, July 4, 1876
Source: Library of Congress, Washington, DC

On the Boston Common, there were "pyrotechnics of Bengal light" and "red fire" in hues from carnation to crimson. The tower of Brattle Church on Commonwealth Avenue, the Town Hall, the Custom House, and other buildings also had displays. In the South End, Chester Park had "an immense number of Chinese lanterns" and more "red fire."[69] And on Bunker Hill, where the day's celebration had started, another calcium light cut the darkness. The Bunker Hill Celebration was one of a series of American Revolution commemorations that celebrated the birth of the nation and sought to reunite it after the Civil War. Boston made a point of inviting southern guests to the event, to underscore their shared national history. As one newspaper declared, "From the 17th of June celebration of 1875 will be dated a new era of peace and prosperity for the once more United States."[70]

A year later, to celebrate the centennial of US independence, New York organized an event that combined thousands of oil lamps, Chinese lanterns, gas jets, and a few blazing electric arc lights at the Western Union Building.[71] It included "processions, artillery, fireworks, illuminations, decorations, and all that the grace of oratory and the charm of poetry can bring."[72] More than one hundred thousand people witnessed the parade of twenty-five thousand, including "torch-bearing temperance societies, firemen, and organizations of mechanics and tradesmen [who] swept into" Union Square, "inundating it with an oppressive flood of light," followed by a roaring rendition of "Hail Columbia" to brass band accompaniment and "thunderous cheers from the vast crowd." Enthusiasm rose to the highest pitch with the singing of the national anthem. Every major building in the city was brightly illuminated, including a huge transparency of the Statue of Liberty on a Madison Square clubhouse. The Union League Club displayed an enormous American flag and many other flags, punctuated by "Venetian lanterns hung suspended from the windows and cornices." The hotels were covered with lanterns, and many buildings used gas to spectacular effect. There was good order, even though the streets were crowded with celebrants. Businesses remained open, and fireworks continued long into the night. Americans considered it one of the most lavish illuminations ever held—longer, brighter, and more exuberant than anything seen in Paris, London, or Rome. The *New York Times* judged that "it was in great degree spontaneous" and "in very great degree due to private citizens." It had exceeded every expectation and "was indeed a night of nights, a splendid festival stolen from dreamland or from the Arabian Nights, whose parallel we are not likely to see again."[73]

Crowds also gathered to see the spectacular lighting of landmarks and natural sites. In *A Tramp Abroad*, Twain noted that the Giessbach waterfall near Interlaken, Switzerland, was illuminated every night. He also witnessed the illumination of Heidelberg Castle. To see it, he slogged through rain on a road "densely packed with carriages and foot-passengers … and finally took up a position in an unsheltered beer-garden." With thousands of others, he watched as "sheaves of varicolored rockets were vomited skyward out of the black throats of the Castle towers, accompanied by a thundering crash of sound, and instantly every detail of the prodigious ruin stood revealed against the mountainside and glowing with an almost intolerable splendor of fire and color. … For a while the whole region about us seemed as bright as day."[74]

Such spectacles were hardly limited to Europe. In 1860, during a visit by the Prince of Wales, Niagara Falls was illuminated. "About 200 colored and white calcium, volcanic and torpedo lights were placed along the banks above and below the American Falls, on the road down the bank of the Canadian side of the gorge and behind the water of the Horseshoe Falls." Bengal lights, normally used as a signal at sea, were employed, "along with rockets, spinning wheels and other fireworks."[75] The event was widely reported and received an enthusiastic notice in the *London Times*.[76] The falls were sporadically illuminated in the following decades, and electricity was first tried in 1881, using sixteen Brush arc lights of 2,000 candlepower.[77] In 1882, an English traveler thought it presented "one of the great wonders of the world in a new aspect. As coloured glasses are now and then placed in front of some of the lights, effects are produced which cannot be witnessed without exciting admiration."[78] These dramatic effects continued during the 1890s and were improved on for the 1901 Pan-American Exposition in Buffalo.[79]

In 1907, General Electric's chief illuminating engineer W. D'Arcy Ryan blasted the cataracts with 1,115,000,000 candlepower from three batteries of projectors. Such an enormous crowd gathered the first night that the suspension bridge swayed perceptibly under their weight and rail traffic was not permitted. Despite this public enthusiasm, many journalists wondered if lighting the falls might not be a desecration more than an enhancement. When the press went out on a balcony to see the lights come on, though, "for a long moment no one spoke a word." Their reports were uniformly enthusiastic. The *New York Tribune* declared that "magnificently illuminated, the Falls were of a beauty that their daylight aspect never equaled. … Every hue in the spectrum was used, and words fail to describe the magnificence." The searchlights equipped with colored filters created a rainbow of effects. The *New York Evening Post* emphasized the awestruck crowd, which "witnessed in dead silence," and "simply gazed and wondered and admired, speechless." The *New York World* predicted that the Niagara illumination would become one of the wonders of the world, for it gave the falls "a new glory."[80] There was an even more intense illumination in 1925, when Ryan projected twice as much light as in 1907.[81] The lighting of Niagara eventually inspired similar treatment of other sites, including the Natural Bridge of Virginia and Old Faithful geyser in Yellowstone National Park.[82] These illuminations further whetted the public appetite for spectacular lighting.

If many illuminations were held in the United States, they were hardly a fading tradition in Europe. Hamburg enthusiastically staged an illumination to welcome the German kaiser in 1895, and in 1906, Madrid celebrated a royal marriage with electrical illuminations and fireworks.[83] The British held an enormous illumination to celebrate Queen Victoria's golden jubilee in 1897. All along the streets were "flaring gas, glow-worm oil lamps, opal globes, paper lanterns and transparencies, incandescent lamps, celluloid flowers, and hundreds of devices in thousands of colored crystals. Everywhere was brilliancy, sparkle, color, [and] at many points a dazzling radiancy."[84] London held another illumination three years later to celebrate victory in the Boer War, and yet another for the coronation in 1902.[85]

Illuminations had evolved from the royal ones of the Renaissance to incorporate new forms of lighting on both sides of the Atlantic. In Britain at the end of the Victorian era, illuminations no longer had "the capacity to agitate social division" and were used "to consolidate collective perceptual experience. Unpredictable and volatile illuminated demonstrations had been replaced by a stylized rhetoric of light."[86] Public lighting, once the perquisite of kings and later a weapon of class warfare, became central to the organization of urban space. It drew attention to a site, defined its contours, increased its importance, and gave it new attributes. Once ephemeral, it had become permanent. If lighting was gradually depoliticized in Britain, in the United States it was becoming more commercialized, intense, and political.

2

Energy Transitions

During the nineteenth century, two public lighting technologies competed for dominance. Gas lighting developed commercially in Britain at the start of the century, and then spread to its colonies, the rest of Europe, and the United States. The gas came not from wells but instead from coal brought to each city, manufactured into gas and delivered to customers by underground pipes. There were gas lighting systems in every major British city by the 1820s, and they added customers for the next half-century. Gas also spread rapidly in the United States. By 1853, New York City's two gas companies had built 246 miles of gas mains that served businesses, private residences, and 9,000 street lamps, but much of the expanding city was not yet served.[1] By 1875, just before electric lighting became commercially feasible, there were more than 400 US city gasworks.[2] These systems were expensive to install, requiring street excavations as well as a central station and coal storage areas.

Modeled in good part on the gas system, the electrical network developed from workable arc lights and generators in circa 1877 to the magnificent incandescent lighting displays of the Panama-Pacific Exposition in 1915.[3] In these forty years, the electrical industry emerged and was consolidated into a few manufacturing firms, notably Westinghouse and General Electric, serving city utilities that were mostly local monopolies. This story is usually told as the movement from gas to electrical lighting, simple to complex installations, and arc lights to incandescent ones.[4] Gas lighting, challenged by electricity, improved to be brighter and cheaper in 1910 than it had been in 1875. Nevertheless, arc lights gradually prevailed, only to be overtaken by improved incandescent lights. This technical story may seem inevitable in retrospect, since compared to gas, electricity released no sulfur, ammonia, or carbonic acid into the air, and consumed no oxygen. Electricity provided a steadier light, and did not tarnish metals, blacken ceilings, increase the humidity, produce unwanted heat in summer, or pose the potential danger of suffocation.[5] Yet gas was not quickly replaced, and one can tell a different tale about how both lighting systems were used to define public space. In that story, the systems were not always in competition but rather were combined to take advantage of their different properties. It must be emphasized that despite much rhetoric to the contrary, neither gas nor electricity transformed night into day. The entertainments and dissipations of the night persisted in the brighter urban geography. As the historian Peter Baldwin concluded, "The codes of behavior that prevailed in the dark streets of preindustrial America proved remarkably resilient, preserving the night as an incompletely civilized realm within the modern city."[6]

The United States passed through several energy transitions before it adopted gas or electricity.[7] Colonial society functioned on the basis of muscle power and small-scale waterpower, and these remained important sources of energy well into the nineteenth century. There were only two steam engines in the United States in 1776, and steam did not become the dominant energy in US manufacturing until 1870.[8] It required fifty years for the energy transition from waterwheels (and water turbines) to steam engines. It likewise required fifty years for gas to spread widely in US cities and towns, and another half-century for electricity to replace gas. During these transitions, cities experimented with different systems. They tried powerful lights placed high above the streets, weaker lights on poles, many kinds of arc lights, Welsbach gas mantles, incandescent bulbs, and hybrid systems of gas and electricity. They also experimented with when artificial lights went on and off, or whether they were needed at all on moonlit nights. In the 1880s, no one could foresee which system would prevail. Experts disagreed even in 1900, and cities had quite-different systems.

Energy transitions require infrastructural change. Coal was essential to manufacture gas or electricity, but it only reached locations served by ships, canals, and railroads. The new transportation arteries established what Christopher Jones has called "landscapes of intensification," where energy transitions accelerated.[9] Sites with ready access to energy sources increased in wealth and population, in a self-reinforcing process. Cities grew and extended their dominance to a wider hinterland. Even within cities, transitions did not occur at the same rate in all sectors of the economy. For decades, gas was primarily sold to commercial and industrial customers, and some urban areas remained without service. Similarly, when electrification began in the late 1870s with the lighting of streets, public buildings, department stores, and hotels, service was concentrated in business and industrial areas. In the 1890s, streetcars adopted electrical motors, but most industries only adopted them after 1900. More gradually, electric wires entered the home, and most rural electrification occurred only after 1935. Public lighting was the first, highly visible phase of a decades-long process.[10]

Energy transitions are not a simple matter of substitution. A steam engine can replace a waterwheel, but a steam-driven factory need not be near falling water. That factory can locate wherever coal can be delivered, and towns that lacked waterpower but could get coal by rail or water became attractive for manufacturing.[11] Thomas Alva Edison modeled his lighting system on the gas system, but that did not mean it would be used for the same purposes. A new energy source favors new locations, undermines familiar living patterns, and offers unexpected possibilities. The railroad could go where the canal could not. The steam factory, unlike a water-driven mill, did not have to shut down during a freeze or drought. Electric lights were far less a fire hazard than gas flames, and could be used in situations that gas could not. Electric streetcars were cleaner, warmer, and threefold faster than horsecars, opening up new suburban areas and bringing electric light to customers along their lines. As these examples suggest, each energy transition makes possible new manufacturing locations, new forms of transportation, different working conditions, alternative settlement patterns, innovations in entertainment, and new living arrangements. People use new energies to create new structures of experience.

Energy transitions take place in stages and only achieve what Thomas Hughes termed "technological momentum" after roughly two decades. This is in part because new energy systems are usually difficult to understand and costly to build. Even a decade or two after their introduction, they may be more expensive than the older system they seek to replace. Therefore, they are adopted slowly.[12] People have habitual ways of building, heating, cooking, and entertaining, and their way of life is embedded in an older energy system that already has technological momentum. In *Networks of Power*, Hughes discerned five stages in an energy transition. The first is the invention of a new system and building prototypes to test it in a few locations. For gas, this occurred between 1798 and 1805, while for electric arc lighting this first stage was from 1875 to 1882. After these initial demonstrations comes the transfer of the new technologies to large markets, notably Manchester and London in the case of gas, and European and US cities in the case of electrification between 1882 and 1890. Next these technical systems grew and achieved economies of scale, stimulating the infrastructures necessary to support the system, such as technical education, factory production, marketing, advertising, and customer service. Only after these three stages did gas or electrification enter the fourth stage of technological momentum. Finally, in the mature stage, each system spread more rapidly than in its first decades.[13]

In Britain, William Murdoch developed gas lighting in 1798 while superintending steam engines in Cornwall. His Birmingham employer adopted the invention and lighted the Soho Factory in 1802. After this successful demonstration, leading manufacturers, including Matthew Bolton and James Watt, promoted gas lighting, and it attracted investors. On London's Pall Mall, gas first was demonstrated in the gardens of Carlton House in 1807.[14] It proved popular, and London built almost three hundred miles of gas mains by 1820. The innovation spread to "nearly every town in Britain with a population over 10,000 by 1826."[15] As Hughes suggests, gas required a quarter-century to reach technological momentum, and even then service had not yet reached most of the domestic market, which grew rapidly during the next quarter-century. British gas technology was transferred to France, Belgium, the Netherlands, Germany, and Scandinavia, often through direct British investment.[16]

Some possible social effects of the new gas lighting were suggested in "A Peep at the Gas Lights in Pall Mall" (see figure 2.1). In this image, people comment on the new gaslights. One expounds on the process of

2.1 A Peep at the Gas Lights in Pall Mall, 1809
Source: Wikipedia Commons

producing the gas, leading his friend to suggest that if the process makes water into a substance one can burn, then the Thames River might "burn down" with "all the pretty little herrings and whales burnt to cinders." A tourist marvels that "we have nothing like it in our country," but a clergyman replies, "Aye Friend, but it is all Vanity. What is this to the immortal light?" A prostitute complains, "If this light is not put a stop to, we must give up our business."[17] None considers the environmental problems that gas production caused, as effluents were flushed into rivers, killing fish and vitiating drinking water. Many voiced these concerns by midcentury, however.[18]

Not everyone immediately embraced gas lighting. As Matthew Beaumont summarizes, "Compared to the intimate forms of illumination associated with the candle and oil-lamp, which lit small areas with an uneven, gently flickering flame, and which consequently generated a kind of contemplative aura, the light from gas lamps seems to its immediate contemporaries distinctly impersonal" and even alienating.[19] Some people objected to artificial light coming through their windows and exposing them to view.[20] In 1816, a newspaper in Cologne listed other objections. Artificial lighting interfered with the divine order of darkness and daylight. Burning gas produced noxious fumes that were unhealthy. Lighting encouraged people to remain outside in the cold, which could increase illness and threaten public health. Moreover, if people no longer feared darkness, then drunkenness and depravity might increase. Possibly, too, thieves would be bolder in a lighted city. There were economic objections as well. Taxpayers who did not want municipal lighting would be forced to pay for it nonetheless. Finally, the *Cologne Zeitung* warned, continual illumination of the streets would "rob festive illuminations of their charm."[21] Even as late as the 1830s, many Philadelphians also were reluctant to adopt gaslight and campaigned against it. They feared explosions, noxious odors, and pollution of the rivers, which would kill "immense shoals of shad, herring, and other fish."[22] A city committee investigated, and found both insurance companies and city officials in Boston, New York, and Baltimore enthusiastic about gas lighting. There was no evidence that it caused more fires. In fact, when the roof of the Boston gasworks caught on fire, the gas flowed without mishap to customers even when the building was enveloped in flames.[23]

Gas lighting was one of the first urban networks. To make gas from coal required a central plant, large storage tanks, and a system of underground pipes to carry the gas to streetlights, public buildings, businesses, and homes. Gas transformed the city, making it brighter and more navigable at night as well as more interesting to look at. Once major streets had gas lamps, more people ventured out, and theaters, restaurants, and various forms of entertainment flourished. The custom of talking a walk at the end of the day became more common, and the French invented the word *flannêur* to describe a fashionable person sauntering through the city streets.

Walter Benjamin famously described the flannêur as "one who walks aimlessly through the streets for a long time. With every step the walking itself gains greater power; the temptations of shops, bistros and smiling women grow less and less, while more and more irresistible becomes the magnetism of the next street corner, a distant mass of greenery, or the name of a street."[24] The flannêurs seldom wandered through the slums or along unpaved streets at the edge of town; they sauntered through well-lighted boulevards and parks, observing café society. They preferred the large cities, such as Paris, Berlin, London, and New York, where their promenading became an end in itself. A large city in the evening hours provided fascinating scenes and encounters. One observer of Paris noted that at night, the city was simplified into a glowing full-scale map, "a web of flames with countless fine threads," in which the flannêur saw "the big and small streets, filaments of light; the open squares, powerful nodes of flame, which shoot their shining rays in all directions—and all around, like a reddish belt of flames, the suburbs." This was Paris "when its 40,000 gas flames burn—Paris by lamp-light," where the walker found "every one of her temptations more tempting yet."[25] In New York, Walt Whitman might be counted among these figures. The flannêur also strolled into the center of a literary genre, the urban sketch. As William Chapman Sharpe notes, the flannêur combined the roles of "reporter, moralist, social critic, and man-about town," and was "a valuable mediator in the urbanite's effort to handle the flux of signs, sights, and social formations that define the modern city."[26]

The flannêur has been a popular figure in cultural studies, but historians have found little evidence that there were many of them. Chris

Otter concluded that the flannêur was at best a marginal figure and that their "dominance is more evident in the late-twentieth century culture studies texts than on the streets of the nineteenth-century city."[27] Joachim Schlor rightly observed that "there have been a multitude of other types of urban walkers; the more varied the city and the streets themselves, the more different walkers (and in time women walkers too) they have engendered. It makes no sense to try and bring all of these forms together under the concept of the *flannêur*."[28] As a thriving business in guidebooks explained, the urban street offered the public unexpected encounters between morality and immorality, law and disorder, ostentatious wealth and abject poverty.[29] There were so many reasons for going out that the solitary flannêur faded into the crowd.

Before 1820, London was thought to be brightly illuminated, but it did not keep pace with foreign rivals.[30] Charles Dickens, who often went for rambles at night, thought "London is shabby in contrast with New York, with Boston, with Philadelphia. … [I]t can rarely fail to be a disappointing piece of shabbiness to a stranger from any of those places."[31] London was also less lavish in its lighting than cities on the Continent. In 1831, London was brighter than Paris, which had only seventy gas jets, placed on a few central streets. But according to one (perhaps-exaggerated) report, "Less than thirty years later, there were more than 100,000 gas jets in the streets" of Paris, concentrated in commercial and wealthy areas.[32] Dickens was struck by the contrast: "The meanness of Regent Street, set against the great line of Boulevards in Paris, is as striking as the abortive ugliness of Trafalgar Square, set against the gallant beauty of the Place de la Concorde. London is shabby by daylight, and shabbier by gaslight. No Englishman knows what gaslight is, until he sees the Rue de Rivoli and the Palais Royal after dark."[33]

Yet the intensity of illumination was not the only aspect to be considered. Many British people agreed with Robert Louis Stevenson, who preferred the warm color spectrum of gas. Stevenson celebrated gas's obvious advantages over oil lamps or carrying a lantern. With gaslight, "a new age had begun for sociality and corporate pleasure-seeking," and "supper parties were no longer at the mercy of a few miles of sea-fog; sundown no longer emptied the promenade; and the day was lengthened out to every man's fancy. The city-folk had stars of their own; biddable, domesticated stars."[34] Stevenson understood that an electric lighting system could be turned on all at once, dispensing with the lamplighter. But like many Europeans, he disliked the glare of arc lighting, which he described as "a new sort of urban star" that "shines out nightly, horrible, unearthly, obnoxious to the human eye; a lamp for a nightmare! Such a light as this should shine only on murders and public crime, or along the corridors of lunatic asylums, a horror to heighten horror. To look at it only once is to fall in love with gas, which gives a warm domestic radiance fit to eat by."[35] By the time Stevenson published his essay in 1878, gas had become traditional, and was associated with the familiar routine of lamp lighting, the ruddy glow of a fire, and domestic comfort. The level of gaslight that seemed brilliant in 1820 appeared either romantic or quaint by the late nineteenth century, when "the visual field produced by the average London streetlamp remained mottled and grainy."[36]

In 1816, Baltimore was the first US city to adopt gas lighting. It made gas from pine tar, but soon switched to coal, following English practice. That same year, Philadelphia boasted the world's first theater with a gaslit stage, though the city did not adopt gas street lighting for two decades.[37] Boston and New York soon followed Baltimore, and by 1827 New York's Broadway was famous for its gas flares. Yet not until midcentury were Gotham's gaslights as numerous as oil lamps, which remained on many side streets.[38] Wealthy homes became important customers, and from their windows spilled out additional light, making their neighborhoods brighter.[39]

Almost all gas was manufactured from coal, and tons of coal were shipped by canal from eastern Pennsylvania's anthracite mines to coastal cities. Pittsburgh's bituminous coal supplied towns on the Ohio and Mississippi Rivers, notably Cincinnati, Louisville, and Saint Louis. Commercial gasworks spread throughout the country. Their first customers were large institutions such as railway stations, factories, city lighting utilities, and a few wealthy homes. Service focused on large customers during the first decades. As late as 1850, Philadelphia had only ten thousand domestic customers, but had hooked up four times as many a decade later. The same accelerated diffusion occurred again with electric lighting, which added most of its domestic customers after 1915.[40]

Street electrification began with arc lights. This required permission from city councils, usually obtained despite opposition from the gas industry. It also required that entrepreneurs or a municipality raise considerable capital, build a generating plant, install street lighting poles, lay underground cables (or string overhead wires), train a staff that could maintain the system, and market the new service to customers who had little knowledge of electricity. Only after an arc lighting system was widely installed on streets and squares, in large stores and railway yards, and at public venues such as parks and roller skating rinks did the electrical company realize a profit. During the months required to install generators, transmission lines, and arc lights, a utility had no return on a large investment.

One of the first demonstrations of electric lighting was in Saint Petersburg in 1872, where Aleksandr Lodygin showed that a carbon filament sealed inside a glass bulb from which oxygen had been removed could cast a brilliant light.[41] However, Lodygin lacked the technical and financial support required to perfect and manufacture his invention. He joined the company of his fellow Russian, Pavel Yablochkov, who developed the electric arc light in Paris, and installed arc lights in the Gare du Nord in 1875.[42] In the United States, several Wallace Farmer arc lights were erected at Philadelphia's Centennial Exposition of 1876.[43] Brush arc lights were demonstrated in Cleveland, and a few were installed in large indoor spaces such as Wannamaker's Philadelphia department store and a newspaper office in San Francisco. But the most famous early use of arc lights was Paris, where "several of the wider streets and squares and about forty workshops" were lighted, including "the avenue leading to the Opera House" (see figure 2.2). As US newspapers reported, this light transformed the night:

> The lamps are placed on posts, precisely like the gas lamps, except that the posts are taller and wider spaced. The lamps are enclosed in large opal glass globes. … As daylight fades away, there comes without warning a sudden flash, and every light in the street is burning with an intense white glare. The effect is like daylight, except in intensity. Every part of the street, the immense traffic in the roadway and the people

2.2 Yablochkov candle arc lights, Avenue of Opera, Paris, 1878
Source: Wikipedia Commons

on the walks, and every architectural detail of the buildings from the top of the roofs, every object however minute in the windows, the flowers on the balconies, are plainly visible and in their natural colors. … Every sign on wall or omnibus, the minute patterns in fabrics and the finest print, can plainly be seen. People seated before the cafes read their papers by the aid of lights on the opposite side of the way. … Every stone in the road is plainly visible and the horses move swiftly along as if confident of their footing. Neighboring streets, though more brilliantly lighted with gas than any American streets, appear dark and gloomy by contrast. … It is impossible to look at the light for more than a few seconds. This intensity, and the occasional flickering of the light, are raised as objections to the electric light. On the other hand, why should anyone look at the lamps any more than at the sun, and when not looking directly at the light, the flickering is hardly noticeable.[44]

The remarkable increase in visibility made it possible to see and do things outdoors at night that once were confined to daytime, such as reading at an outdoor café or closely examining the pattern of a passing woman's dress. Where gaslight provided a narrow range of muted colors, the arc light replaced this sepia landscape with one where fabrics and flowers retained their daylight appearance. Gaslights had made the city recognizable and navigable, but electricity made possible a wide range of nocturnal behavior once only possible during the day.

The transformation was considered so desirable that central cities paid a premium for arc lights, which almost always proved more expensive than gas. In London, arc lights were erected on the Embankment in December, 1878, where they attracted curious crowds (see figure 2.3).

Just before Christmas in 1880, Brush arc lights went up in New York "like stars emerging from the darkness." From Fourteenth to Thirty-Fourth Streets, they first appeared as "a row of small white dots scattered along the street, growing in size each instant until, after the lapse of a few minutes, they had developed into large powerful electric jets, burning somewhat irregularly at first, but soon settling down into a white steady glare."[45] By 1881, Brush arc lights had also been adopted on the wharves of Montreal, in London's Charing Cross and Paddington Stations, and in Royal Albert Hall.[46] The hall's "vast interior was wholly flooded with a steady, white light" equivalent to thirty thousand candles. The "colors of pictures on the walls and of ladies' dresses were revealed with the clearness of daylight. All persons present but the holders of gas stock were charmed with the effects."[47] Wolfgang Schivelbusch incorrectly declares that "in quality, gaslight and electric light were almost interchangeable"—a statement that might be defended after the perfection of the Welsbach gas mantle, but no one who viewed the electric arc systems in 1881 in Paddington Station or the Royal Albert Hall would have agreed.[48] Schivelbusch underestimates the complex choices that the various lighting systems presented, overlooks the centrality of arc lighting for a quarter-century after 1877, and seems unaware of the differences between US and European choices. He tends to assume that the industrialization of light was a rather-uniform phenomenon in which cultural differences were less important than universal effects.[49]

2.3 Yablochkov candle arc lights on the Victoria Embankment, London, 1878
Source: Edward W.C. Arnold Collection, New York Public Library

The light cast by gas or the arc light was not similar to Edison's incandescent bulb. His lamps were brighter and whiter than gaslight, and briefly seemed to pose a threat to the glaring arc lights. William Hammer demonstrated the Edison system in London, where he installed 230 lamps on the Holborn Viaduct in January 1882 as well as a successful exhibition at the Crystal Palace.[50] London thus had the first Edison central station, even before New York. London simultaneously allowed three companies (Thomson-Houston, Lontin, and Siemens) to put up 33 arc lights each for a one-year trial. A government report found, however, that "the value of the excess of light" was dubious; "for public thoroughfares, uniform distribution of light, generally speaking, best meets the public requirements, and this can be most successfully obtained by many small lights at small distances apart. Powerful centres of light at long intervals apart give intense brightness with deep shadows in their immediate vicinity, and distribute the light very unequally over the areas assigned to them; tested by this principle the excess of light given by the electric [arc] lamps was much less valuable than might be supposed."[51] Rather than opt for one of the arc lighting systems, the report praised the Holborn Viaduct, where Hammer had placed "two incandescent lamps in each Gas lantern, each lamp, … giving about the same light as an ordinary Gas lamp, and the two therefore double the light of the gas lamp disused. No lamp is more than 66 feet distant from another. There is scarcely any part of the Viaduct which is better lighted than another; there are no strong shadows to deceive the eye and the footstep; there is no flickering and no material variation in illuminating power. … [T]he Viaduct is, for all practical purposes, well lighted."[52] The Edison station there had a capacity of 3,000 lamps, most of which were installed in shops and hotels, lighting up the entire quarter.[53] When the service expanded to supply 400 incandescent lights to the telegraph office, it briefly seemed that Edison's system might soon outcompete gas indoors and arc lights outdoors.[54]

In these years, each system required its own generating plant. For example, the larger train stations used electricity to light the yards, the repair shops, waiting rooms, and platforms. Strasbourg's station had an isolated plant that supplied both arc and incandescent lights.[55] Standalone systems were also installed on ships and in wealthy homes.[56] Generating plants were small by modern standards. The largest Edison central stations, in New York and Chicago, could supply 15,000 lamps. The systems installed in smaller cities and towns usually could light up only 2,000 or 3,000 lamps.[57]

In both Britain and the United States, gas systems long remained larger than the emerging electrical network. Arc lights were expensive. By 1885, New York had replaced 3,016 gaslights that cost $52,780 a year to operate with 647 brighter arc lights that cost more than three times as much: $165,308.[58] Most streets still had gas lamps. The arc light spread more slowly in Britain, where there were only about 700 in the entire kingdom in 1890.[59] More arc lights were installed in Paris, but as an economy measure late at night these were extinguished and gas jets lighted.[60] Nor was this practice unique to Paris. In 1911, a highway engineer from Boston toured European cities to study their public lighting. He found that in London and Berlin, high-pressure gas lighting was quite effective and compared well with electric arc lights. He also found that to economize, "in many of the European cities it is the custom to locate gas lamps between the electric light lamps" that were "extinguished at midnight, at which time the gas lamps are lighted."[61]

US gas companies thwarted electrical competitors by lowering prices and improving their technology. In 1878, 1,000 feet of gas had cost $2.50 in Cleveland and $2.75 in Baltimore. A decade later the price in both cities had fallen to $1, and similar price cuts occurred in New York, Philadelphia, and Chicago.[62] Some gas companies lowered or eliminated the cost of hooking up for service. A few fought electric competitors by using political connections, as in Hartford, Connecticut, where the superintendent of the gas company sat on the Common Council. In Hartford, "malignant personal attacks were made by the gas manager upon the American [electric company] officers." Such tactics only succeeded in the short run, however, and by 1884 Hartford was rapidly shifting away from gas lighting.[63] Improved technology defended the gas interests more effectively than slandering the competition. In the 1890s, gas firms aggressively sold the new Welsbach gas mantle, which dramatically increased the light produced. The Welsbach mantle was based on the realization that while burning gas itself gave off more heat than light, the high temperatures could make other substances incandescent. Invented in Europe in 1885, the mantle contained oxides of thorium and cerium that when heated, cast a brilliant white light. So equipped, a gaslight produced six times more light. Its adoption prolonged gas systems into the twentieth century. In the United States, more than 10 million Welsbach burners were in use by 1900.[64]

Gas companies also benefited from favorable long-term contracts. In Chicago, for example, one gas company monopolized service based on an "all-embracing blanket-franchise." After 1900, when the city and state legislature attempted some control, they met fierce resistance. By 1906, Chicago's official statistician reported that "the present relation of the City of Chicago to the People's Gas Light and Coke Company is in every way not only unpleasant, but amounts to veritable warfare." New York struggled with similar problems, and again the state legislature became involved in regulating a powerful local monopoly that had considerable technological momentum.[65] After seventy-five years of development, gas was built into the infrastructure of major cities; it was less expensive and brighter than ever before, and enjoyed legal protections.

Nevertheless, arc lights increased their market share. How did they work? Each had two carbon rods, with a small gap between them (see page 249). The electric current jumped across this gap, and the illumination was emitted "due to the intense heat of the tips of the carbon rods," which became white hot, and "to a smaller degree to the arc itself."[66] During the quarter-century when arc lights were most common, they were often powered by direct current (DC). The positively charged upper carbon rod was larger, and became much hotter and produced most of the light. Because it was bombarded by negatively charged particles, it gradually acquired "a hollow center" called the "crater of the arc" that directed the light downward to where it was most wanted. The smaller negatively charged rod remained cooler, burned down more slowly, and became pointed. Changing to alternating current (AC) arc lights presented some difficulties. In AC arc lights, the rods were alternately positive and negative, and both became pointed. The temperature remained lower than in a DC arc light, no crater formed, and the illumination was shed equally in all directions. AC arc lights therefore required a reflector to direct the light downward more than DC arc lights did. Regardless of the current used, the arc light was extremely hot, burning at fifty-five hundred to six thousand degrees Fahrenheit, and produced a light too powerful to look at directly. Because it was so bright, the arc light usually was hung higher above the street than gas lighting. It required more maintenance than modern streetlights, because the carbon rods burned down rapidly and had to be replaced—a task performed in the daytime. The positive rod burned down about twice as fast as the negatively charged one, and an important innovation was a mechanism based on the electromagnet that automatically advanced the rods as they burned down, keeping the gap between them constant. Arc lights required less maintenance than gas lamps, which for decades had to be serviced by a lamplighter every day. By the 1890s, though, gas lighting was being automated, using an electric sparking system.

When the first street arc lights were erected, some businesses paid to be linked into the system. In Boston, Elihu Thompson recalled that "the stores along the streets would frequently have one or two arc lights installed within them, and these would be operated from the same light circuit in series with the street lights."[67] Arc lights had definite advantages for use in large indoor spaces. Because they did not produce smoke or acidic vapors, they were readily adopted by clothing and department stores. "Arc-lighting systems were relatively simple to install and operate." They had "almost no accessories or extras like fuses, meters, switches, switchboards, and so forth." Improved arc lights gradually became cost-competitive with gas. By 1891, the Thompson Houston Electric Company proudly announced in *Western Electrician* that it had installed precisely 87,131 arc lights. A decade later there were 400,000 installed, and the price for each fixture had fallen to $15.[68] Most of these replaced gas lighting.

The arc light shed a strong light compared to the far weaker and more reddish gaslight, but had drawbacks. At first it required a great deal of current. The intensity could not be adjusted, unlike a gas flame or kerosene lamp that could be turned up or down. The arc light's extreme temperatures also rendered it dangerous. Accordingly, many inventors sought an alternative that was enclosed, incandescent, cooler, longer lasting, and lower maintenance. The arc light forced an electrical current to jump across a gap between carbon rods. In contrast, an incandescent bulb had a hair-thin filament with a high resistance that heated up and glowed when electricity passed through it. To prevent the filament from burning up, air was evacuated from the bulb or it was filled with an inert gas. Several inventors were trying to invent something along these lines, but Edison's laboratory invented not only a bulb but also an entire practical system, including power generation, wiring, switches, sockets, and every other detail. The Edison bulb was well suited to domestic use, but initially was not powerful enough to supplant the arc light out of doors. The arc light was still the standard street installation in 1904 when General Electric's leading scientist Charles Steinmetz improved it with a magnetite electrode that had a high melting point.[69] His innovation was brighter, and the new electrodes burned for 600 hours, or every night for roughly two months. Improved carbon rods had lasted only 125 hours.

Among the many improvements to incandescent bulbs, the most important was a tungsten filament, with a melting point over sixty-one hundred degrees Fahrenheit. Created at General Electric's research laboratories, its color spectrum shifted away from infrared to higher frequencies (see page 251). The tungsten light more closely approximated daylight. It began to replace carbon filament lamps after 1909, followed by the gas-filled tungsten arc light in 1914. Powerful lights with tungsten filaments quickly started to supplant arc lights. Nationally, between 1913 and 1917 incandescent streetlights doubled to almost 1.4 million, while the number of arc lights fell by 25 percent to a quarter-million.[70] By 1922, gas lighting was a relic, and arc lights had been largely replaced, although there were still 12,882 on New York's streets.[71] Yet the gas industry prospered, and the gas and electric utilities became intertwined. Already in 1899, 40 percent of US gas companies also sold electricity, and many electrical utilities had acquired gas companies.[72] The two energy forms began to serve different markets. In 1880, gas was almost exclusively for illumination, but fully 80 percent of sales were for cooking and heating by 1919. A decade later, gas was scarcely used for lighting, even though annual sales were higher than ever.[73]

Edison's first central station, on New York's Pearl Street, supplied incandescent lighting to businesses, wealthy residences, and corporate clients, such as the Stock Exchange, but gas remained a formidable competitor, as domestic customers were accustomed to it, and the gas system was literally built into their architecture. Over 1,000 miles of gas mains beneath New York's streets supplied factories, theaters, office buildings, and homes as well as streetlights. In 1892, these streets had 27,000 gas lamps and only 1,199 arc lights.[74] Gas also persisted in Boston, Baltimore, and Philadelphia, although less so than in London and Paris. Because gaslight was dominant, both Charles F. Brush and Edison modeled their lighting distribution systems on gas lines, and focused on how to deliver light as cheaply as gas did (see figure 2.4).[75] Philadelphia provides a snapshot of how electric arc lights were introduced. It had installed 561 arc lights by 1885, trying out "the Brush, Thomson-Houston, Fort Wayne Jenney and United States systems." These 2,200-candlepower lights burned "all night and every night" at a cost of "$200 per year" for each light, for a total of $112,200 per year.[76] This was considered too expensive. Arc lights worked well, but most of Philadelphia was still lighted by gas.

After 1875, the simultaneous use of several kinds of illumination was common. Even in the first decade of the twentieth century "the nocturnal landscape … was a crazy quilt of different forms" of illumination that included tower lighting, arc lights, gasoline or kerosene lights, gaslights, and incandescents.[77] Chicago had more than 38,000 lights installed in public places in 1913. Half of these were not electric: 12,700 gas lamps and 6,500 gasoline lamps. The other half consisted of 17,000 electric arc lights and 2,050 tungsten incandescent lights.[78] In other words, just before World War I, in the second-largest city in the United States—a city that had mostly been rebuilt after the great fire of 1871—there were 4 different public lighting technologies in use, and just a single light in 20 was an incandescent bulb. Nor was Chicago unusual in this regard. Most cities were slow to adopt incandescent street lights, and retained arc lights and gas systems well into the twentieth century (see figure 2.5). This persistence strongly suggests that the new technologies themselves did not drive the redefinition of urban space, but that the process was inflected by social, aesthetic, and political factors as well as economic considerations.

As late as 1910, the US electrical system supplied less than 20 percent of factory power and lighted no more than 10 percent of the nation's houses. Only in 1925 had electrification become the norm for US manufacturing and urban living.[79] A smaller percentage of the British or Germans then lived in electrified homes, in part because they had a longer acquaintance with and investment in gas technologies.[80] Robert Fri has noted "the far-reaching societal transition that must accompany

Notes from old Edison Note book.
(Edison's writing)

1

The fact to be accomplished by me is the invention perfection and introduction into practice of a practicable system of illuminating by electricity which shall effect every object and take the place of the present method of lighting by gas

That the cost of any light or any number of lights equal in candle power to gas is to be but one third that of the latter,

That the cost of plant shall be no greater than gas

2.4 Thomas Edison, Notes on Electric Light and Gas
Source: Hammer Papers, National Museum of American History, Smithsonian Institution Archives, Washington, DC; photograph by David E. Nye

2.5 Servicing an Arc Light
Source: Hall of History, Schenectady, NY

transformation of the physical energy system." This larger transformation is unavoidably slow because "the energy system is not simply a collection of autonomous pieces of plug-and-play technology. Rather, it is an integral part of our individual lives, influencing where we live and shop, shaping how we establish social networks, and molding countless everyday habits."[81] The United States installed an electrical infrastructure somewhat more rapidly than Europe, but it nevertheless required half a century, and even longer if one includes rural electrification.[82]

The diversity in lighting during the slow transition to electricity is beautifully illustrated in 1900 by the experience of a committee of five who were sent from Cincinnati to visit ten other cities and evaluate their street lighting systems.[83] By 1875, Cincinnati had a well-established gas lighting system with "5,290 public lamps connected by 170 miles of supply pipes."[84] Owned privately and worth $6 million, it expanded to serve more domestic customers and had great technological momentum by 1900. Yet the city was ready to upgrade to either an improved gas system or one of the new arc lighting systems. The committee first went to Saint Louis and saw 936 DC, enclosed arc lights concentrated in the business district. The enclosed arc gave a steadier light at a lower cost than the open arc light, and its carbons needed replacing after a week or so.[85] Saint Louis had not adopted AC, however, even though it cost less when transmitting power over long distances. The Saint Louis public preferred the electric light, but to expand service, the city needed to install an AC power plant, redo all the wiring, and purchase AC arc lights. Instead, the city purchased new Welsbach mantles for 11,930 gas lamps.[86] These mantles gave six times more light without burning any more gas, and their operating cost was only one-third that of an arc light.

The committee next went to Indianapolis, which they found to be behind Saint Louis. It had no Welsbach gas lamps and used "the old open type of arc lamps."[87] They left almost immediately for Chicago, which planned "to replace all of the gas lamps." In contrast to Saint Louis, lighting one mile of a Chicago street electrically cost only 10 percent more than using gas.[88] Yet nearby Milwaukee calculated its gaslights cost one-fifth as much as arc lights. While most of its 2,400 gas lamps were of the older type, Milwaukee had 100 Welsbach lamps on trial, and seemed likely to adopt them. At this point, after visiting four cities, the Cincinnati committee had found that only Chicago definitely preferred electricity, but still had a preponderance of gas. Saint Louis and Milwaukee had upgraded gas systems, and Indianapolis was too old-fashioned to teach them anything.

The committee next found that Pittsburgh had a preference for electric lights, even though it had access to locally produced, inexpensive, natural gas that was far cleaner than coal gas.[89] Half of the city's 2,572 arc lamps were "of the open type," and the other half "of the enclosed type," which were much preferred because they cast fainter shadows, glared less, and could be "hung lower avoiding the foliage." They also had fewer outages and threw light out "to a greater distance."[90] Pittsburgh had 3,150 "open-flame gasoline burners" in outlying areas and new districts where streets had not yet been paved. The goal, as in Chicago, was to replace them all with arc lamps. The open type were "placed approximately 350 feet apart" and "hung on the middle of the streets, about thirty feet high." The enclosed arc lamps were "on ornamental goose-neck fixtures," had "poles at the curb-line," and were "eighteen feet from the surface of the street."[91] The committee realized that Pittsburgh, being the headquarters of Westinghouse, might favor electric lighting in order to support local industry. Therefore, they also visited the suburb of Edgewood, which had 110 Welsbach mantles burning natural gas at an annual cost of just $22 each. They "found that the lighting from these incandescent gas lamps gave good satisfaction, as the light extended to a great distance."[92]

In Pittsburgh, cost clearly favored a gas system. Yet the committee had also established that an arc light was so bright, it replaced three or more gas lamps—a calculation Baltimore officials confirmed. The choice lay between gas lamps equipped with the Welsbach mantle or enclosed arc lights using AC. The latter, when covered by an "opal inner globe" that diffused the light agreeably, could be placed much closer to the ground than the older arc lights to provide bright, shadowless illumination without glare.[93] The problem was that these two systems were seldom in use side by side, where they might be compared in identical atmospheric conditions. By this point on their journey, some cities had nothing to teach them, notably Philadelphia, whose technologies were old—23,000 ordinary gaslights and 8,348 open arc lights. New York had many different lighting systems, but presented a confused picture due to the power of gas monopolies as well as conflicts between the utilities and local government. But Boston was unambiguous. It had extensive experience with the latest technologies and had chosen the enclosed arc lights instead of Welsbach gas mantles. Hartford had already shifted entirely to such arc lights, but the committee thought these were placed too far apart.

In Washington, DC, the committee finally found the leading alternatives arrayed side by side. At the "Thomas Circle your committee had an exceptionally good opportunity of making a comparison of the different systems of lighting from a single point, as there are six streets radiating out from Thomas Circle. … On one street there are enclosed-arc lamps; on another street open-arc lamps; on a third street were open-flame gas lamps; two streets were lighted with Welsbach gas lamps."[94] From this vantage point, under identical atmospheric conditions, the committee saw that enclosed arc lamps gave the brightest and most even distribution of light. They "could see a quarter of a mile down this street, and distinguish at any point an object" on the sidewalk. The Welsbach gas lamps were not quite as good, but had "a spectacular effect" and "satisfactorily illuminated the surface of the street," where objects could also be clearly distinguished. The committee decided that in residential neighborhoods, "the glare of the open arcs" might be "objectionable," which suggested that either enclosed arc lights or the less expensive gas lamps might be used there. (On stormy nights, however, many gaslights blew out.) A Washington official assured the Cincinnati committee that "a considerable number of Welsbach gas lamps" would be installed in the residential districts even though along important streets arc lights were replacing gas.[95]

In its report, the committee recommended that Cincinnati adopt the enclosed AC arc light with a frosted inner globe that reduced glare and diffused the light evenly. "Another advantage of the alternating-current type of lamp is that lamps of 1,200-candlepower can be used, if desired, from the same dynamo that furnishes the 2,000-candlepower lamps." This meant that ten lights instead of six could be erected "at the same cost per year for lighting, enabling the lamps to be placed much closer together. The use of an increased number of lamps" provided "the same amount of light more evenly distributed."[96] Choosing electricity over gas might seem obvious in retrospect, but it did not seem so when London and Saint Louis preferred gas. In 1900, it was necessary to send out a committee to inspect lighting systems and confer with local experts. Cincinnati's committee found that even in 1900, there were more gas-lights in the largest US cities than electric arc lights.

One must distinguish between the date of an invention and its widespread use. Between 1875 and 1915, public electric lighting spread at a moderate pace, and the adoption of domestic electric lighting was slower still.[97] Adoption of a new system of energy is not a sudden event but occurs gradually. Sir Humphrey Davy demonstrated an arc light at the Royal Institution in 1808, but this did not mean London's streets were soon electrically illuminated. Davy used 2,000 voltaic cells to produce enough current, but this was not an economical way to generate and distribute power for a far-flung system. It took seven decades of research and development before it was feasible to adopt arc lights instead of gas. Edison's demonstration of his enclosed, incandescent light in Menlo Park, New Jersey, in 1879 was a milestone, yet for the next twenty-five years, few Americans had it at home and either gas or arc lights provided most public lighting. Even so, between 1880 and 1915 Americans adopted electric lighting more rapidly than most of Europe, for reasons explained in the following chapter.

3

The United States and Europe

Between 1870 and 1915 the US city experienced almost convulsive growth. Not only did millions of immigrants arrive, not only did millions more leave the countryside for the growing metropolis, and not only did vast new industries emerge but also the very fabric of cities changed, including a shift in scale from horizontal to vertical, an expansion into far-flung suburbs, the emergence of giant department stores, and the invention of new forms of entertainment such as professional baseball, amusement parks, movie theaters, and dance halls. Through all these transformations, like a brilliant shaft of light, ran the electrification of the city. Electricity drove the streetcars that carried passengers from suburbs into city centers. It carried people in skyscraper elevators, department store escalators, and subways, making possible the immense concentration of humanity at the urban core. Electricity lighted up the city after dark, operated the rides at amusement parks, projected films, amplified music, and made new spaces potentially useful. Using electricity, anywhere could be lighted or ventilated, whether a sports field at night, a basement, an attic, the roof of a skyscraper, an underground corridor, or a tunnel beneath the streets. Electricity was an enabling technology that Americans and Europeans used to reshape the fundamental nature of their cities.

Well before electric lighting was woven into the urban fabric, the networked city emerged.[1] Historians trace the origins of these networks to Roman roads and aqueducts, or sewers built in the Renaissance, but such networks expanded rapidly during the nineteenth century, including water, sewage, gas, the telegraph, pneumatic tubes, burglar alarms, police call boxes, and fire alarms. The concept of the modern city included a wide range of networks by the time electric street lighting became a practical possibility. Electrification did not merely add one more system but also linked the many networks together. Electric traffic lights regulated the flow of traffic. Electric thermostats controlled the temperature in houses, offices, greenhouses, and food processing plants. Electrical timers regulated industrial ovens. Electrical bells warned of fires, announced a visitor, signaled the end of an event, or indicated the arrival of an elevator. The elevator's movements could be monitored electrically, as could the flow of liquids in pipes, water levels in reservoirs, or number of people passing through subway turnstiles. Electricity made possible a vast network of networks, tying the city together. It facilitated complex transportation systems, rapid communication between skyscrapers, simultaneous transmission of prices for stocks and commodities across the country, and many forms of illumination.

A growing public traversed the new, public sphere that had expanded along with the networked city, with its water supply, gas lighting, fire alarms, police call boxes, sewer systems, and other services. These created a zone of safety, cleanliness, and comfort that attracted people into the streets, and made it more convenient to enjoy concerts, lectures, theaters, restaurants, cafés, or other pubic activities. This public not only saw the city transformed by fog, mist, or rain, at dawn, noon, and twilight, but also saw each site transformed as lighting multiplied perspectives. When wet, an illuminated city's streets became distorted reflecting mirrors; when covered with snow, they were brighter than in soft summer darkness. This protean city was by definition heterotopian, constantly offering new experiences in familiar locations. The constant play of new impressions fascinated writers and artists, who taught readers how to see, negotiate, and enjoy these new urban scenes. Likewise, women were determined to enjoy more than chaperoned promenades on the newly lighted streets.

Electrical devices also furthered oversight and control, and became part of the police apparatus in both Europe and the United States. Alarm systems, augmented by the telephone system in the 1880s, made it easier to coordinate activities from police stations. The electric light, like gaslight before it, was widely understood as a means of crime prevention. Yet if the city became brighter, during the final decades of the nineteenth century in New York, Paris, London, and Berlin, there was rising public concern about "the danger, insecurity, and immorality of the nocturnal city."[2]

Because electrification served so many needs and linked the networks of city services, it became a symbol of progress and was considered an essential part of any modern city. Good public lighting became a highly visible indicator of modernity, and across Europe visitors from smaller cities realized that they had fallen behind the newly electrified centers of London, Paris, and Berlin. Yet these metropolises were behind the United States. Initially in Britain, the incandescent "light was ridiculed as a Yankee notion of your Mr. Edison. Nobody could be got to invest a cent in the business."[3] British investors only began to take an interest in 1882. A member of the British Institution of Electrical Engineers admitted that "notwithstanding that our countrymen have been among the first in inventive genius in electrical science, its development in the United Kingdom is in a backward condition, as compared with other countries, in respect of practical application to the industrial and social requirements of the nation."[4] Even as late as 1896, Piccadilly Circus had only a few, simple electric signs and was a dim contrast to New York's Broadway (see figure 3.1).

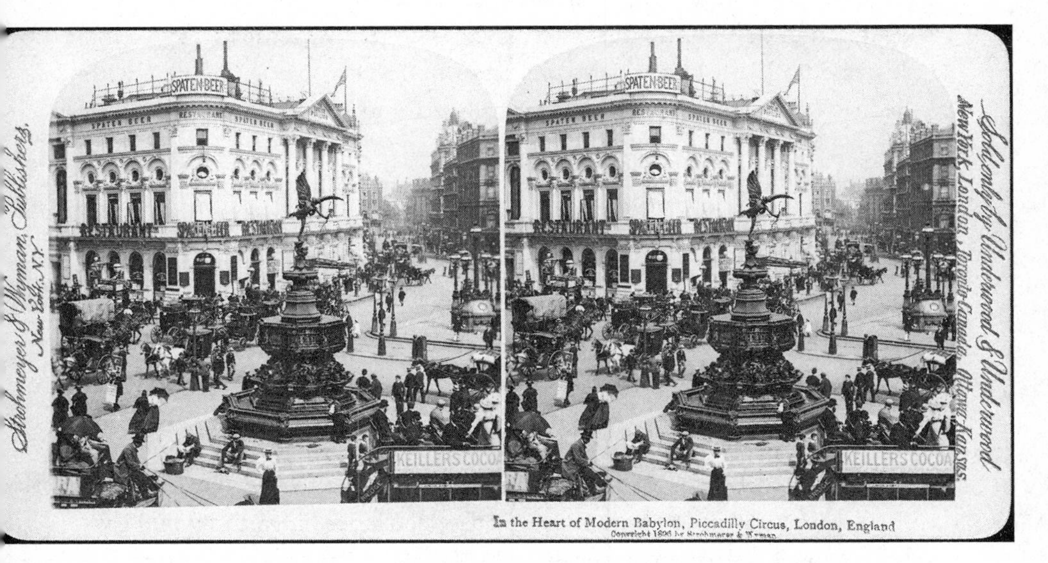

3.1 "In the Heart of Modern Babylon, Piccadilly Circus, London, England, 1896"
Source: New York Metropolitan Museum of Art

As William Preece, a leading British electrical engineer, declared in 1884, "I know nothing more dismal than to be transplanted from the brilliantly illuminated avenues of New York to the dull and dark streets of London. This happened to me very recently." He had departed from a New York hotel "to the Cunard Wharf, a distance of about four miles, through streets entirely lighted by electricity." When he arrived in London, he "drove from Euston to Waterloo without seeing a single electric light." While in North America he had also been in Montreal, Buffalo, Chicago, Saint Louis, Indianapolis, and Philadelphia, "finding in each city the principle streets and warehouses, as well as stores and places of public resort, lighted by arc lamps. After so much brilliance in these cities, the return to dull gas has a most depressing influence." Preece did criticize US lighting, however, noting that the arc lights were "usually fixed on much taller posts than we are accustomed to see in England, arranged in zigzag fashion along the streets, at a distance from each other of about fifty yards." The overall effect was "brilliant" but "by no means perfect, and no effort seems to have been made to distribute the light uniformly, as has been done in England." Preece concluded that the problem of lighting "does not seem to have been attacked at all from a scientific point of view." Instead, the "commercial spirit alone appears to have been exercised in developing this enterprise."[5] The British actor Henry Irving described a New Jersey railroad station in 1884 as being "lighted with electric [arc] lamps, which occasionally fiz and splutter, and once in a while go out altogether. Nobody pays any attention to this. Everybody is used to the eccentricities of the new and beautiful light."[6]

A German electricity expert flatly declared in 1885 that "electric lighting for streets and public places is more used in America than in Europe," even though the Europeans installed technically excellent systems, such as that on Leipziger Strasse in Berlin. He concluded that "the greater use of electric lighting for streets in America" was not due to any technical superiority.[7] By the early twentieth century, the electric illumination of US streets was far more intensive than in Europe (see figure 3.2). In Liverpool, for example, which some regarded as Britain's best-lighted city in 1901, there were 152 electric arc lights and 9,000 gaslights.[8] In all of Britain, fewer than 20,000 arc lights were on city streets, and gaslight predominated.[9] By contrast, in 1903 the greater New York

3.2 GE Parade Float, ca. 1900
Source: Hall of History, Schenectady, NY

City area alone had 35,000 arc lights, and a decade later, 19,000 arc lights and 18,000 tungsten incandescent streetlights, then rare in Britain.[10]

In 1903, a quarter-century after the commercial introduction of arc lighting, the *Electrical World and Engineer* contrasted the illumination of US and European cities. On both sides of the Atlantic, the domestic use of electricity had grown slowly, and most of the market was for lighting in public spaces and workplaces, but there the similarity ended. The editors calculated for large cities the equivalent of the number of 16-candle-power lamps per 1,000 inhabitants, and found Boston was far and away the most brightly lighted. The five leading US cities had more than three times more illumination of Paris, London, or Vienna.[11]

Even Saint Louis, which relied primarily on gaslight, was far brighter than Paris. European cities were not dark but they were dim.

Table 3.1
Intensity of Public Lighting, European and US Cities in 1903

City	16-candlepower lamps per 1,000 inhabitants
Boston	1,232
New York	859
Chicago	730
San Francisco	660
Saint Louis	600
Paris	185
London	184
Berlin	176

Source: "Electric Lighting in Boston," *Electrical World and Engineer*, September 19, 1903.

One British journal declared of English cities in 1895, "The principal streets are lighted in a manner which astonishes the foreigner and incites the American to contemptuous scorn."[12] In the same year, a British study found that only "about 3 percent of the total urban artificial light" in Britain was electrical.[13]

Why was there such a disparity between Europe (especially Britain) and the United States? There are at least five conceivable explanations. For starters, European and US cities might have had access to different electrical technologies. Second, gas might have been decidedly cheaper than electricity in Europe. Third, European and US cities had contrasting street layouts, one of which might have been more favorable to infrastructure development. Fourth, gas might have had greater technological momentum in Europe than in some parts of the United States. And finally, cultural preferences might have played a role: Did many Europeans simply prefer gaslight?

It is difficult to make a case for a technical explanation. During the last quarter of the nineteenth century there was a free exchange of scientific information through international congresses and publications, and there was also a continual migration and remigration of electrical specialists across the Atlantic. In the 1880s, Edison and his US rivals, George Westinghouse and Thomson-Houston, moved quickly into foreign markets. Moreover, many of the skilled people who developed the

US electric lighting system were European immigrants. Some of Edison's foreign employees returned home in the 1880s to start utilities and electrical equipment manufacturers. Francis Jehl returned to Austria-Hungary and played an important role in its electrification, and also sent continual reports back to the United States.[14] Thomas B. Thrige went home to Denmark and built up an electrical manufacturing company. S. J. Bergmann went back to Germany, where he eventually employed thirty thousand people in electrical industries.[15] There was also much movement in the other direction. The German Charles Steinmetz emigrated to the United States and became the most famous researcher at General Electric. Nicola Tesla emigrated from Serbia, worked briefly for Edison, and played a key part in developing AC for Westinghouse.[16]

Early on, Thomson-Houston sent agents to Europe, bringing its international holdings to the merger that formed General Electric in 1892. Westinghouse was early in entering the British market, and the companies it later spun off remained two of the three largest there in 1930.[17] In short, it is simply not plausible to explain the slower development of electric street lighting in Europe by arguing that Europeans lacked access to technical information or expertise. Between 1880 and 1914, US corporations participated in every important British or French exposition, and European corporations likewise exhibited in the United States.

International General Electric held considerable stock in European lamp-producing firms as well as substantial daughter companies in Austria, Belgium, France, Great Britain, Germany, Greece, Holland, Hungary, Italy, Portugal, Spain, and Switzerland.[18] These interests grew, and by 1931, 25 percent of General Electric's total equity was held in Europe.[19] In France, where it had invested only half as much as in England or Germany, by 1922 just one General Electric subsidiary, Cie Francaise Thomson-Houston, employed ninety-five hundred people in nine factories. Another General Electric subsidiary, the Compagnie des Lampes, was France's largest producer of incandescent lamps.[20] US electrical companies in some cases had seats on boards of ostensibly competing firms.[21]

Symptomatically, the first electric sign spelling out a word was produced by an American, William Hammer, but was displayed at London's Crystal Palace in 1881. He also invented the first flashing sign, seen at a Berlin exposition in 1883 (see figure 3.3). Since multinational enterprises

3.3 First Flashing Electric Sign, Berlin
Source: Hall of History, Schenectady, NY

exported the full array of electrical technologies in both directions across the Atlantic, it seems clear that not technological differences but instead cultural factors shaped which technologies were selected and how they were used.[22] US and European cities had access to the same lighting technologies, supplied by the same international corporations. Differences in urban lighting did not express differences in technical knowledge. In Europe, public lighting more often emphasized public buildings and historical monuments. There were fewer billboards or flashing advertising signs. A German engineer who visited the United States in 1913 criticized the US practice of street lighting. He preferred municipal to private utilities, and told readers of the *American City*, "In Germany the best practice in street illumination is to have powerful flaming arc lamps swung at some height rather than to have [as in the United States] more numerous and smaller lamps placed at a lower elevation."[23] He argued that arc lights should be thirty to sixty feet above the street to secure an even, cost-effective illumination. Higher placement also brought out more architectural features of nearby buildings. (He was unaware of earlier, extensive US experiments with tower lighting, which are the subject of the following chapter.) Because US public lighting unevenly alternated between brighter and dimmer areas, he contended, it fatigued a pedestrian's eyes by forcing pupils continually to expand and contract. He also complained that public and private lighting were not coordinated, and thus produced a "jumbled and inartistic" effect.[24]

Yet Americans could be equally critical of European public lighting. Twain remarked that "London is still obscured by gas—gas pretty widely scattered, too, in some of the districts; so widely indeed, that except on moonlight nights it is difficult to find the gas lamps."[25] By the 1890s, the US technical press often noted the disparity between gaslit Britain and the electrified United States. The *Electrical World and Engineer* remarked, "We in America have become so thoroughly familiar with the effectiveness of powerful electric lamps for street lighting that it seems queer to think of the long period that has passed without adequate illumination in the streets of a great metropolis like the city of London."[26]

Was this due to a difference in costs? British coal and labor each were 25 percent less expensive than in the United States, but these lower expenses benefited both gas and electrical utilities. A more crucial difference was that the United Kingdom regulated gas prices, and imposed controls on stock watering and dividends.[27] US gas companies were less regulated and so reaped large profits even in the 1880s after they lowered prices in response to competition from electricity. Another important difference was that British gas companies were able to sell residual products (notably coke) for a higher price than their US counterparts. Yet if US gas was more expensive than British gas, it remained considerably cheaper than US electricity. Price was not decisive. US cities were willing, even eager, to pay a premium for what they regarded as the superior illumination of electricity. British cities were more cost-conscious and retained gas systems a generation longer.

British gas systems also had greater technological momentum than those in the United States. They were established earlier and were more deeply intertwined with the urban fabric. They had been naturalized after generations of use. A wide range of people and institutions had a vested interest in maintaining that technology, including not only those in the gas industry itself but also its suppliers and customers. Gas was widely used not only for street lighting but for interior lighting, heating, and cooking too. In contrast, many US cities had only a brief period of gas service, especially those west of the Mississippi. Kansas City had no gas lighting until after the Civil War; Denver did not begin to construct its gas system until 1870, scarcely a decade before electrical arc lights posed a challenge.[28]

Another cultural difference was the role of government. The national government did not play any important role in promoting US street lighting. Washington, DC, itself did not have an extensive or high-quality lighting system. In 1887, it had eighty-seven public arc lights—fewer than Waterbury, Connecticut, or Watertown, New York.[29] In 1891, it had but 75 percent as many hours of lighting per year as Baltimore, Minneapolis, or Jersey City.[30] With the exception of presidential inaugurations, nineteenth-century Washington did not stage notable civic events on the scale of New York, Chicago, Boston, Philadelphia, Saint Louis, or San Francisco. The driving forces in holding expositions or installing new lighting systems were city councils and local businesspeople. Neither Congress nor state legislatures had much to do with it. Washington never held an important exposition. Paris, London, Berlin, and Vienna held frequent world's fairs, and the European capitals took the lead, whether installing street lighting or staging spectacular events. The United States was the world leader in urban electrification, yet Washington lagged behind other US cities. Instead, city councils, business elites, local utilities, and the major electrical corporations had considerable freedom to experiment with and develop new forms of illumination. US expositions were held in a parade of cities, including Louisville (1883), New Orleans (1884), Chicago (1893), Atlanta (1895), Nashville (1897), Omaha (1898), and Buffalo (1901). These frequent events continuously encouraged a cadre of lighting engineers to improve display techniques. And because expositions were held in all parts of the United States, spectacular forms of lighting became known and were rapidly adopted throughout the country. In contrast, European capitals focused attention on themselves.

European historians have argued that the working class resisted the erection of gaslights and resented their regularization of urban space. Europeans saw public lighting as an extension of the police power and therefore understood it as part of the state apparatus of control. In Germany and France from the late seventeenth century on, resistance to authority often was expressed by smashing lanterns. Schivelbusch notes that during the French Revolution, some representatives of the old order were hung by the neck from lantern posts, and during nineteenth-century uprisings crowds smashed gaslights.[31] In the Paris insurrection of July 1830, crowds attacked the gaslights because they symbolized both the king's surveillance as well as a new social order focused on the boulevards and commercialized space.[32] Politically motivated lantern smashing was far less common in Britain, however, and there were few, if any, such attacks in the United States, where lighting systems had no connotations of federal power but rather were local (often privately owned) symbols of each city's modernity and progress.[33]

Another contrast between the United States and Europe was their different road systems. European main roads frequently converged on a cathedral or palace. This was almost never the case in the United States, where streets usually were laid out in straight lines in a checkerboard pattern. This meant that European networks, whether supplying water, drainage, gas, electricity, or telephone service, often had to work around cathedrals, palaces, and the homes of nobility, which could obstruct efficient delivery of services and force expensive detours in laying cables, stringing wires, or digging tunnels. In contrast, the most powerful class in the United States was composed of successful businesspeople, many of whom invested in new urban networks. By the early twentieth century, successful entrepreneurs like Isaac Merritt Singer and Frank W. Woolworth built skyscrapers, which further intensified land use and projected the networked city skyward.

Most European cities originated in premodern times and had winding streets at their core. Many properties had a narrow street frontage and long back lot. The streets themselves were usually the only public space. As cities grew, vehicles increased in number and grew in size, clogging central districts and creating a demand for broad streets in new areas.[34] Cities also began to set aside parks and squares as public spaces. By the nineteenth century, cities like Barcelona, Copenhagen, and Paris had a warren of crooked narrow streets in their medieval core. Beyond, streets were wider, and blocks more uniform and interspersed with occasional parks.[35] The new streets were not for pedestrians because of traffic intensity. Instead, wide sidewalks offered a new social space that included trees, displays of goods, kiosks, cafés, benches, and even urinals.[36] The nineteenth-century continental boulevard encouraged strolling, window-shopping, and social encounters.

London, in contrast to this pattern, was somewhat less densely populated than Paris, had fewer apartments, more small dwellings with gardens, fewer wide streets, and a more privatized social life.[37] Builders of networks for gas and electricity in London faced the disadvantage of narrower and irregular streets. The Parisian boulevards that Dickens admired were arguably easier to electrify than London's crooked streets; those in Chicago certainly were. Yet even continental boulevards were still darker than major thoroughfares in the United States. In Europe, "it is the rarest thing to find any windows lighted after dark. The shutters are pulled down tight, and the lighting of the streets [was] dependent entirely upon the street lamps."[38] In short, Europe relied more on gaslight with little light from private sources, while US streets were intensely electrified as well as further illuminated by advertising signs and lights from private residences.

Moreover, few European cities grew as rapidly as those in the United States, notably those west of the Mississippi, which experienced explosive growth just when electric lighting became a practical alternative to gas. Minneapolis grew from a population of just 45 in 1845 to 129,000 in 1885. Denver had less than 5,000 people in 1870, but more than 70,000 fifteen years later. Kansas City tripled in size to over 100,000 people between 1870 and 1885. Throughout the nineteenth century, US elites felt a cultural inferiority to Europe, and sought to compensate by building new infrastructure, museums, and universities. An Omaha art museum epitomized the problem. A Danish journalist who visited in the 1890s saw a new building, large enough to house a major collection, with an excellent electric lighting system. But the walls were bare; the museum had not yet acquired any art.[39] Such cities knew they lacked history, art, or tradition, and accordingly claimed not the past but a brilliant future. They emphasized new buildings and state-of-the-art technical systems. Instant communities like San Jose, Denver, or San Francisco had little sentimental attachment to the gas system, and quickly adopted arc lights.

US cities did not have medieval origins and resembled the newer sections of European cities. In New York, the lower end of Manhattan had a premodern street pattern, but most of the city was laid out on the grid system. Shortly after the American Revolution, the national government divided all unsettled lands into squares, based on the astronomical determination of longitude and latitude. When US cities began to grow rapidly in the early nineteenth century, this grid pattern already had been established as the national norm, in contrast to narrow and irregular European streets. Nineteenth-century Paris required an expensive clearing and renovation project to create its famous boulevards. New York's broad avenues and Broadway were laid out before the city expanded. Every US city and town established after 1790 was divided into a checkerboard of streets.[40] This was quite evident in newer cities such as Cincinnati, Cleveland, Chicago, Saint Louis, Omaha, Minneapolis, Denver, and Salt Lake City (see figure 3.4). Only the oldest neighborhoods of New York, Baltimore, and Boston resembled Europe.[41] Rather than being shaped by military considerations, US cities were located and built for largely commercial reasons. Their layout facilitated the development of networks of services, from gas lines, water mains, and sewers, to telephone lines and electrical conduits.

3.4 Aerial View Salt Lake City
Source: Library of Congress, Washington, DC

In addition to all of these factors, by the late nineteenth century, Americans had developed a tradition of celebrating the nation through technological display, which gave lighting a different cultural significance than in Europe. Americans did not merely build the world's largest system of railroads but also opened new lines on Independence Day amid flag-waving crowds. They constructed the world's largest suspension bridges and invented the skyscraper, and considered both sublime.[42] Technology was early woven into the national identity, not least because Americans lacked traditional European rallying points such as a royal family, national church, and ancient historical sites.[43] If the United States could boast no Stonehenge, Roman Coliseum, or ruined Greek temple, in 1915 it could celebrate and extravagantly illuminate the Brooklyn Bridge, Eads Bridge over the Mississippi, Woolworth Building, and other technological achievements. These became iconic, and Americans purchased millions of photographs and postcards that depicted them.

European elites saw little reason to cover a royal palace or national cathedral with light bulbs. Nor did they necessarily see electric light as a superior form of illumination. While Americans were generally enthusiastic about the powerful light cast by electric arc lights, many Europeans agreed with Stevenson's critique of them in favor of gas. As Otter remarks of Victorian Britain, "The flight from yellowness [of gas] was not universally lauded. Most people were accustomed to seeing yellow. This is how normal night appeared: ochreous, cosy, peppery. The whiteness of electric illumination was often an unpleasant shock, registered chromatically as bluish."[44] Preece, who had complained of the lack of electrical light in London, observed that compared to gas, it indeed appeared blue at first. But "Americans did not call it blue at all," and after one became accustomed to it, "the imaginary blueness rapidly disappeared."[45] Yet it was not imaginary. Scientific tests showed that emanations of electric light contained more blue than daylight did. In contrast, light from burning coal gas had—compared to daylight—roughly three times as much red, only 40 percent as much blue, and little green.[46] Stage actors complained that electric light changed the appearance of their traditional makeup and costumes, which had been developed with gaslight in mind. On the other hand, textiles under electric light appeared much more as they did during the day, and drapers in London and department stores in New York enthusiastically adopted Edison's system. In short, in 1900 one could make a case for gas or electric lighting. Recall that the superiority of electric arc lighting was not obvious to the committee from Cincinnati that traveled to ten cities in order to select a new street lighting system. The British preference for gas in 1900 was defensible on aesthetic and economic grounds, even if it seemed odd and backward to Americans. It was a matter of cultural preference.

Visitors to the United States noticed the differences in illumination (see figure 3.5). A well-traveled British visitor to New York encountered its public lighting as something extraordinary and entirely new in 1883,

3.5 New York Street Lighted by Brush Arc Lights, 1880
Source: New York Public Library

when London had few electric lights and Paris was five years away from establishing service (as opposed to small plants serving particular buildings or small areas).

> Looking along Broadway, New York, one sees the bright, white, moonlight effect of electric lamps every here and

> there—here in front of a store, there in front of a theatre, beyond in front of a hotel. Electricity has superseded gas in the interior arrangements of not a few hotels and shops. Within a few weeks Mr. Edison placed about 16,000 lamps in the stores and offices of New York alone. It is, however, in the streets and squares that the new light is most conspicuous. … [T]he effect of the light in the squares of the Empire City can scarcely be described, so weird and so beautiful is it. Enormous standards, rising far above the trees, are erected in the centre of each square. From these standards the light is thrown down upon the trees in such a way as to give them a fairy-like aspect. Except for the temperature, it would be easy to imagine, even on summer nights, that they were covered with hoar frost. Immediately beneath the standards the shadow of every leaf and branch of the interposing trees is imprinted on the asphalt. As the leaves themselves flutter in the passing breeze, the shadows they cast on the pavement below appear very like living objects. Such is the delicate tracery figured on the footways that the pedestrian who makes his first acquaintance with the phenomenon feels almost afraid to walk lest his footsteps should obliterate the lovely handiwork of the electric light.[47]

As he traveled, this visitor realized that New York did not lead but rather followed a national trend. The most brightly lighted city, he thought, was Cleveland. US cities were adopting new forms of lighting more extensively and rapidly than their European counterparts.

The cultural and historical differences between European and US cities partially shaped the technological choices they made. For example, the greater openness and sprawl of US cities pushed utilities to adopt AC. DC was economical within a radius of only about one mile from the power plant. The further DC power was transmitted, the larger the copper wires had to be, and copper was expensive. In densely populated European cities, a one-mile radius could provide a profitable customer base. But in US cities like Chicago, population density fell off rapidly beyond the central business district and a different system was needed.

Westinghouse developed Tesla's AC system in the later 1880s and proved its worth at Chicago's Columbian Exposition. AC could be transmitted longer distances over thinner cables than DC. This in turn meant fewer central stations were needed, and these could install larger, more efficient generating equipment. If Chicago at first had a crazy quilt of small DC generating stations, these soon consolidated into a large AC system under the leadership of Samuel Insull.[48]

In contrast, due to restrictive legislation, London's electrification stagnated in the 1880s, and during a modest expansion in the 1890s it was balkanized into many small firms.[49] As the *Electrician* noted, "The fanatical dread of monopoly has resulted in their being no business to monopolize."[50] Several utilities used high-speed engines originally developed for marine work, coupled with DC generators. These systems were smaller than steam engines and often could be put in existing buildings. It did not pay to extend such DC systems more than one mile from the power source. The many small, DC generating stations of the British utilities could not easily reap the benefits of load sharing.[51] In 1895, in Britain, "about 3 percent of the total urban artificial light" was electrical, and even in 1903 the moon could be the dominant form of illumination (see figure 3.6). In the United States, on the other hand, "by 1899, 60 percent" of urban lighting was electrical. Germany developed AC electrical systems after 1891, and by 1900 was at roughly the same level as Britain.[52] French development was slowed by court rulings that protected gas monopolies until after 1906, while the Germans surged ahead with the creation of regional power networks.[53] In 1911, Berlin and Chicago each had six large power stations, while French cities had "a mosaic of technical systems," and London had 64 small plants. In that same year, Chicago consumed twice as much electricity compared to London, even though it had only one-third of London's population.[54]

A decade later, in 1921, London had a patchwork of 80 electrical services with "50 different systems of supply, operating at 24 different voltages and 10 different frequencies. The same pattern persisted throughout the country."[55] Anyone moving to a new city or even new neighborhood risked having equipment ill adapted to the voltage at the new address. Of the ten leading European cities, London had the largest population (more than Paris and Berlin combined), but the lowest

3.6 London and the Thames River, Night, 1903
Source: Library of Congress, Washington, DC

electrical consumption per capita. The interlinked reasons for London's expensive and inefficient electricity supply included overlong reliance on DC current, competition from a deeply entrenched and well-regulated gas system, misguided legislation that discouraged private investment, and a maze of local political jurisdictions that made load sharing or consolidation difficult.[56] Nevertheless, serving an upper-class clientele, London's utilities profited more per kilowatt-hour sold than in any other large European city.[57]

US urban geography was more suited to electrification than the twisting streets and irregular layout of European cities. It was easier to lay electrical cables on broad, straight streets arranged in a grid pattern. Furthermore, lack of an established church, royal family, or hereditary aristocracy meant that cities were less governed by traditions, and few places were considered off limits to development or display. Most European cities resisted the construction of skyscrapers, preferring that churches remain the tallest structures. In the United States, skyscrapers redefined urban space, projected the grid upward, further concentrated utility customers, and became sites for dramatic lighting. US cities were less inclined than their European counterparts to resist either advertising or commercial illumination, whether on the streets or sides of buildings. Americans made lighting such a central element of the city that one critic declared there was not one form of modernity but two: "The Europeans get the Day, but we get the Night."[58] Not all Europeans were impressed, however. When first shown the lights of Broadway, G. K. Chesterton remarked, "What a glorious garden of wonders this would be to anyone who was lucky enough to be unable to read."[59]

One final crucial difference remains. The United States adopted the automobile far earlier than Europeans. This development came at the end of the period of experimentation with urban lighting systems, and is inseparable from the development of the assembly line, which accelerated the production of US automobiles and lowered their price. By 1915, more than two million cars were sold every year in the United States. In the middle of the 1920s, there was one car for roughly every American family, in contrast to one for every hundred people in Germany.[60] After circa 1915, US cities wanted brighter streets that cars could negotiate at thirty miles an hour. Pedestrians might be satisfied with pools of light from streetlights, but drivers wanted an even, general illumination in an urban environment with reflective surfaces and signs that were visible using headlights. A full generation before their European counterparts, US cities accommodated drivers, changing not only the illumination of the street but also its overall appearance. Americans devoted increasing amounts of urban space to automobiles, which both expanded the area to be illuminated and decreased the number of pedestrians. European cities, more densely populated and with far fewer automobiles, required less illumination. They also restricted electric advertising to a smaller scale. As US cities decentralized, the primary purpose of streets became automotive transportation, and street life still familiar in Europe disappeared, including delivery boys on bicycles, street vendors, sidewalk displays, and children playing on the pavements.[61]

US street lighting provided a visual framework for private illumination, particularly in commercial districts. These additional light sources included both window displays and electric signs hung above them. In theory, these extra light sources might have increased the perception of a harmonious night cityscape, especially if the buildings were similar in architectural style, size, and scale. But in practice the many separate businesses adopted an enormous variety of styles, colors, calligraphy, and scales of presentation. As a result, even where streetlights created a harmonious sense of perspective, this effect was often overwhelmed by the flashing discordance of a hundred businesses each seeking maximum attention.

National Electric Light Association members, who sold urban lighting systems, knew that the placement of the poles along with the arrangement of reflectors and shades not only maximized the brightness of the street but also threw nearby businesses into relative obscurity. At their 1908 meeting in Chicago, they discussed how a new street lighting system, properly installed to focus light on the street and sidewalk, forced store owners to defend themselves by investing in new signs and more intensive show window lighting.[62] Association members also knew that intensifying the illumination of one street drew customers to that area and forced darker streets to adopt equally intensive lights. Competition increased between streets as well as between businesses on each block, driving up the consumption of electricity. In contrast to Europe, where urban utilities were frequently owned by the city, local governments in the United States typically played a small role once they had granted a company permission to operate. They might regulate the size and placement of electric signs so that they did not obstruct the street, but for the most part city councils reacted to problems more than they imposed technical standards or developed an aesthetic.

Rising demand for night commercial lighting had implications for the full realization of the transition from gas to electricity. Utilities profit most if they can even out demand, avoiding high peaks and deep valleys in consumption. Since they cannot store large quantities of electricity, they are most efficient and profitable when demand does not vary too greatly from day to night. They need to "balance the load." In 1905, US utilities wanted to increase demand at night because it fell once industry and businesses closed. When salespeople sold streetlights, show window lighting, and electric advertising, they helped their utilities even out the demand, which meant that less generating equipment lay idle. In the United States, intensive public lighting increased utility profits, but European electrical utilities faced a different market. Gaslight in the United States had all but disappeared by 1910, yet as late as "the 1930s about half of London's streets were lit by gas," and across Britain millions of domestic customers still prepaid for gas via a slot meter.[63] With weak evening demand, British utilities had little incentive to unbalance their load further by increasing sales during the day. Therefore, the contrast between the United States and Europe intensified in the 1920s.

When three General Electric executives traveled through France in 1928 to investigate the lighting industry, they were surprised to find "no street lighting consciousness whatsoever" in smaller towns. "The pedestrians were provided with hand lanterns, and people bidding their guests good night were holding lanterns or lamps to see that they did not stumble" as they stepped down into the street. In Paris, Monte Carlo, Milan, and Venice, they found that shop windows in the evenings were often dark and shuttered "while crowds of people were still passing. ... This entire failure on the part of the shopkeeper to appreciate that light had a drawing power was incomprehensible to an American business man."[64]

Gas systems and the social customs that accompanied them had so much technological momentum in Europe that this factor by itself might seem adequate to explain why the transition to electricity was slower than in the United States. But other factors also played a role, notably the organization of US cities according to the grid pattern, rapid growth of new US cities after 1850, decentralization of political power, and nationalistic identification with spectacular technological displays. Taking these factors together, it would be surprising had US cities not adopted electricity more rapidly and completely than their European counterparts. Indeed, during the 1880s, many new US cities embraced the now-forgotten system of tower lighting.

4

Moonlight Towers

In 1763, a British magazine published a proposal "to light whole urban areas by four oil lamps, each at right angles" that would be "mounted at the tops of high pillars."[1] In 1834, Philadelphia briefly considered erecting a 300-foot tower with a glass enclosed fire on top to light the entire city.[2] And in the 1880s, Paris contemplated installing an 1,100-foot electric tower with a hundred powerful arc lights as a nighttime sun.[3] Burning gas, tar, petroleum, or other substances did not produce enough light to make such towers practical, but powerful electric arc lights made them feasible. Schivelbush briefly noted that Detroit tried out such a system and tower arc lights enjoyed some success in the United States during the 1880s, but these systems have never received much attention, perhaps because most were in smaller towns and new cities.[4] Older centers with established gas systems like Philadelphia or Boston were less likely to invest in something so radically different. But new cities in the Mississippi Valley and West experimented with powerful lights on towers 125 to 250 feet aboveground.[5] These systems had an aesthetic quite distinct from that of either gas or arc lights at the street level.

Tower lighting raised many questions. Should the electrified city appear much as it did under the stars and moon, or might it take on an entirely new look? Should natural light be emulated or surpassed? Gaslight had already extended the length of the day for work, shopping, and amusements, without overwhelming natural illumination from the moon. Should life retain its ancient division between day and night, or should electricity be used to abolish that distinction, ushering in a society that never slept? In the 1920s, a professor argued in the *Literary Digest* that "the alteration of day and night is a check on the freedom of human activity which must go the way of other spatial and temporal checks."[6] Edison thought that human beings might be so affected by electrification that they would require less sleep. As he put it, "In the old days man went up and down with the sun. A million years from now he won't go to bed at all. Really, sleep is an absurdity, a bad habit."[7] He suggested a campaign to urge people to sleep less.

Such proposals were conceivable after two generations of intensified electric lighting. At first, powerful lights had only been presented occasionally at special events. The idea of tower electric lighting was discussed in 1844 in Cincinnati where a young inventor, John W. Starr, developed a prototype electric arc light that he patented in England (patent 10,919). He contended that his light would be so bright that it would be best to mount clusters of them on poles two hundred feet high. He claimed he could illuminate Cincinnati as bright as day.[8] Starr died shortly after receiving his British patent, however, and his light was never developed.

Three decades later, improved dynamos made arc lights commercially feasible, but communities with a well-established gas infrastructure did not rush to invest. Aside from opposition from entrenched gas interests, which lowered their prices to meet this competition, the early arc lights were so much brighter that their glare was hard on the eyes at close range. This intensity indicated that Starr had been right. They might be mounted higher above the streets than gas lamps ever had been and illuminate larger areas. Brush was one of the first to put up arc light towers, notably in New York City. He received permission in 1880 to erect conventional polls along Broadway and Fifth Avenue (see figure 3.5), and higher towers in Union Square and Madison Square.[9] The *Electrical World* declared there were "few outdoor sights more beautiful than Madison or Union Square in summer time, with the rays from the tall electric tower shining down at night on the masses of foliage and the throng of promenaders."[10]

Several companies developed tower lighting systems. In New England, on Nantasket Beach, the Northern Electric Light Company erected three wooden towers one hundred feet high and three hundred feet apart, which had arc lights of ninety thousand candlepower, illuminating a field where two baseball teams played what may have been the first night game. "The design of the exhibition was to afford a model of the plan contemplated for lighting cities from overhead in vast areas, the estimate being that four towers to a square mile of area" would be sufficient, though the reporter covering the event was doubtful.[11] Tower lighting did not catch on in New England, but the Jenney Electric Company of Indiana installed tower lights in many western cities and smaller towns. The Thompson-Houston Corporation's arc lights were also used for this purpose in cities such as Council Bluffs, Iowa, and Mobile, Alabama.[12] In 1888, Thompson-Houston purchased a majority of the Jenney stock, which became part of General Electric interests in 1892.[13] General Electric gave technical support to existing tower systems, but did not promote new ones. Yet neither this consolidation of electrical manufacturers nor the public preference for more conventional lighting standards was foreseeable in the early 1880s, when tower lighting briefly seemed the system of the future.

Brush drew national attention to tower lighting with a spectacular early installation in Wabash, Indiana.[14] A small community, Wabash could be lighted by four arc lights placed on the city hall's flagstaff. On March 8, 1880, a dark and drizzling night was dramatically transformed before a somewhat-skeptical crowd of ten thousand people. They included members of nineteen city councils, curious to see if it really was possible to light up an entire community using arc lights. This event took place less than three months after Edison had drawn many reporters and a curious crowd to see his incandescent lighting system displayed in Menlo Park.[15] Electric lighting was still so new that most people had never seen it. A journalist described the event, witnessed by a crowd that knew only gaslight:

> Suddenly from the towering dome of the courthouse burst a flood of light, which under ordinary circumstances would have caused a shout of rejoicing from the thousands who had been crowding and jostling each other in the evening's darkness. No shout or token of joy, however, disturbed the deep silence that suddenly settled on the vast crowd. … The people, almost with bated breath, stood overwhelmed with awe as if they were near a supernatural presence. The strange, weird light, exceeded in power only by the sun, yet mild as

> moonlight, rendered the courthouse square as light as midday. … Men fell on their knees, groans were uttered at the sight, and many were dumb with amazement.[16]

The crowd's awed reaction was a classic example of the technological sublime, in which people are literally struck dumb when the power of the object they have come to see vastly exceeds their expectations. Yet as people became accustomed to electric lighting, such a display began to seem unexceptional, then natural, and within a generation, inadequate. Lighting engineers discovered that to hold the public interest, what seemed sublime today had to be outdone tomorrow.[17] After this moment of amazement, the crowd started shouting and cheering. Reporters wrote extravagant headlines such as "Wabash Enjoys the Distinction of Being the Only City in the World Entirely Lighted by Electricity" or "The Entire City Brilliantly Lighted and Shadows Cast at Midnight Five Miles Away." In the aftermath, many predicted remarkable effects from the powerful lighting. It might keep the chickens up all night until they died of exhaustion. It might stimulate the growth of nearby crops. It might encourage people to work night and day.[18] *Scientific American* reported the experiment, but did not endorse such predictions.

The Wabash demonstration inspired other communities to install such a system. The Jenney company had a particularly effective salesperson, Ronald T. McDonald, who had a private railcar, and "would roll into town with a flourish, hire a band, and start a parade to draw the crowd to a public hall," where he would sell them on tower lighting. By the time he left, he often had a contract.[19] Aside from its spectacular appeal, tower arc lighting had practical advantages. It cost less to establish, because it required fewer lights and less wiring than arc lights on short poles spread along the major streets. Putting in such conventional installations also disrupted traffic more than erecting a few central towers. In small towns, moreover, a tower system's dynamo was usually in a nearby basement or building, which meant all parts of the system were close together, simplifying supervision and maintenance. By 1887, Jenney was selling a double carbon lamp; "when one of the carbons burns out, the current immediately brings the other carbon into circuit, and the light" remains on.[20] This innovation cut in half the service visits to each tower (see figure 4.1).

Jenney and Brush were promoting an entirely new function for electric lighting. Rather than restrict it to the street, they envisioned a general transformation of the night. They provided light equivalent to what people were accustomed to on a cloudless night under a full moon. As one promoter put it, "There is little doubt that, except in large cities built of solid blocks of high buildings, a satisfactory lighting can be obtained by the tower system with a smaller number of lamps than by any other. By satisfactory is meant, not a brilliant lighting of the center of the city with the residence portion and outskirts in darkness, but a general and nearly uniform lighting of the whole area sufficient to enable persons to walk and drive [a horse-drawn vehicle] comfortably."[21] By

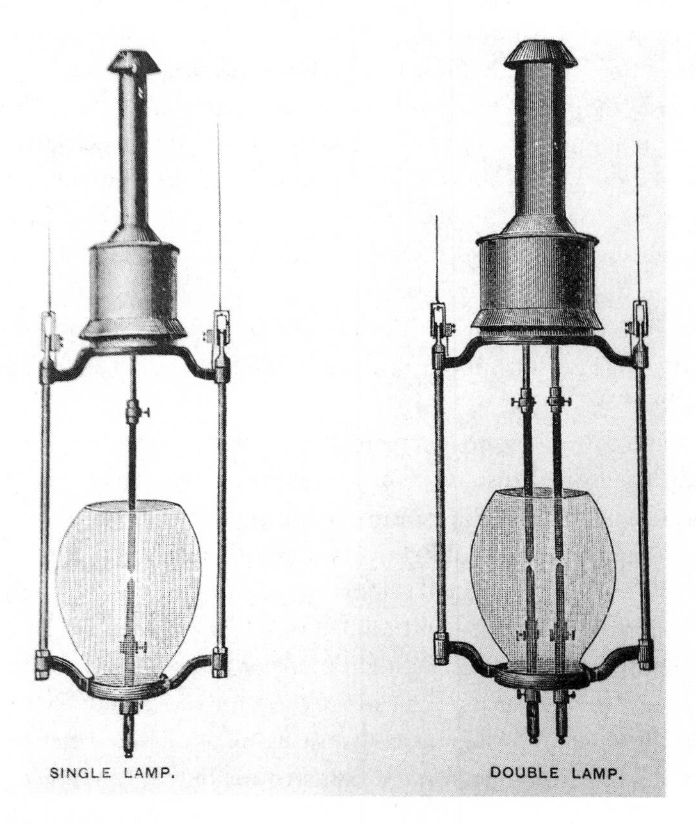

4.1 Jenney Arc Lights
Source: 1887 Trade Catalog, Jenney Light Company, Smithsonian Institution, National Museum of American History Library, Washington, DC

the late 1880s, such tower systems could be found in much of the Midwest, including Council Bluffs and Davenport, Iowa; Detroit and Saginaw, Michigan; Eau Claire, Fond Du Lac, and La Crosse, Wisconsin; and Elgin, Rock Island, Peoria, and Evansville, Illinois. (see figure 4.2). The inauguration of these systems was described enthusiastically. A reporter in Elgin, for example, called it a "day of jubilee" when twenty-four towers made the town "radiant 24 hours a day."[22] Many Indiana communities adopted tower lighting, no doubt because the Jenney company was located in Fort Wayne, Indiana. Cities in the South and West also adopted tower lighting, including Louisville, Kansas City, Fargo, Denver, Portland, Stockton, San Francisco, San Jose, Los Angeles, San Diego, Austin, Atlanta, Mobile, Savannah, Ashville, Houston, and New Orleans. Cities frequently turned their systems off during a full moon when additional light seemed superfluous.[23] San Francisco and Denver, for instance, employed "the moonlight schedule for all night lighting."[24] Moonlight seemed an appropriate standard. In Minneapolis, too, an "electric moon" was placed on a 257-foot tower.[25] As late as 1912, in a brochure explaining conventional streetlights, General Electric declared that "it is by no means necessary that the intensity of illumination be great. Rather it should be uniform. The full moon casts no very bright light; yet it illuminates the earth so uniformly that the impression of soft brightness is produced. The full moon, not the blazing sun, is to be emulated in street-lighting."[26]

Most tower systems were in the United States, but the City of London tried out Brush's system in 1881. The British journal *Engineering* wrote approvingly, "There is something exceptionally fascinating and attractive in the thought of lighting a great city by a number of powerful lights suspended in midair far above its roofs, shedding a purely white light, softened by distance and robbed of its glare by height; and strangers arriving in London by the night trains or boats—for the river would be rendered navigable by night as well as by day—could not fail to be deeply impressed."[27] After thirty Brush lights illuminated the area from Blackfriars to Cheapside, it seemed the tower system might catch on in Britain.

The entire city of Detroit installed a Brush tower system in 1881. It first erected ninety towers. Twenty in the central business district were

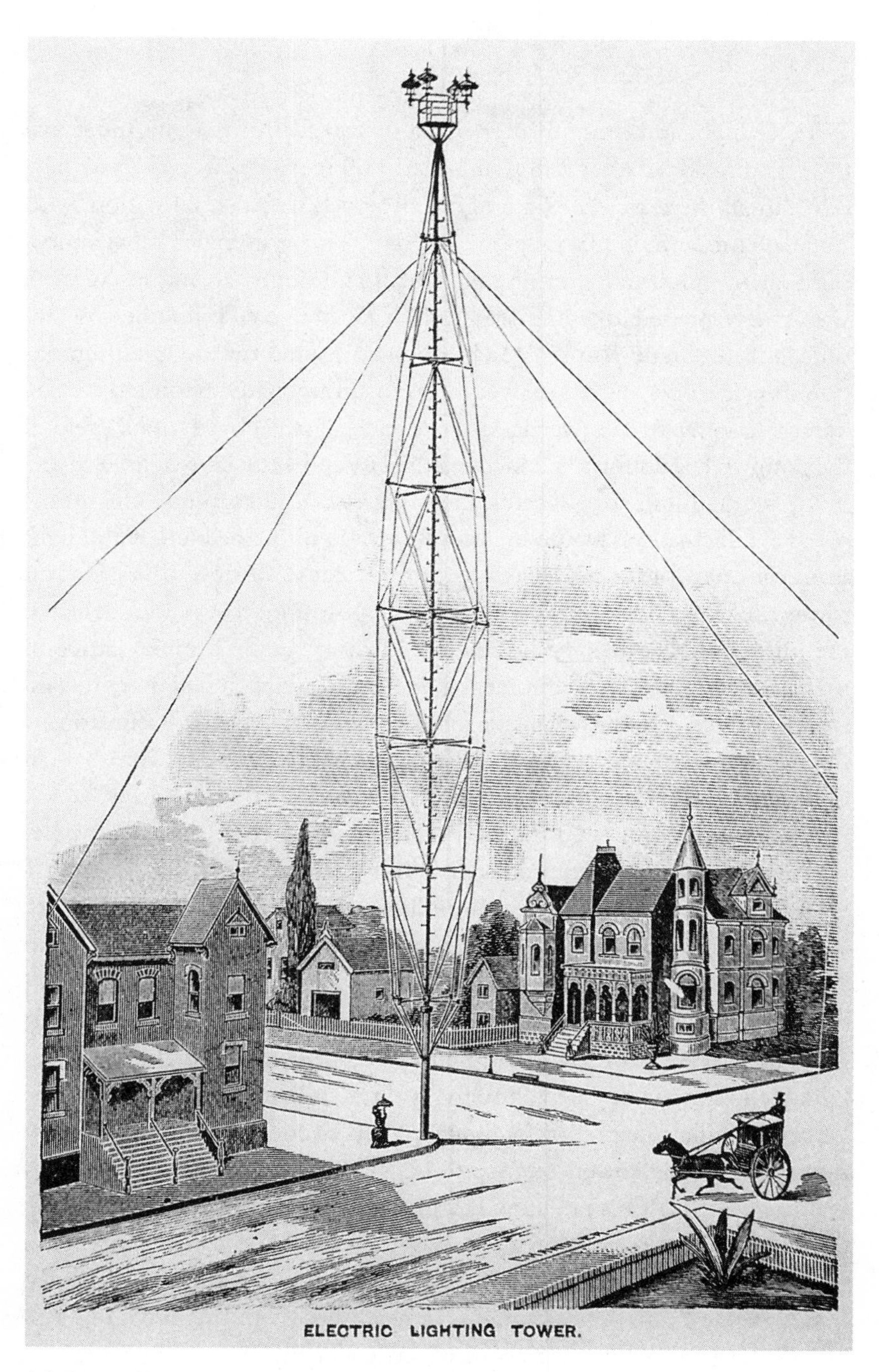

4.2 Jenney Tower
Source: 1887 Trade Catalog, Jenney Light Company, Smithsonian Institution, National Museum of American History Library, Washington, DC

175 feet high, and those further out rose 150 feet in the air. Additional towers were added after 1883, laid out 1,000 feet apart in triangular patterns. In all there were 382 arc lights, all served by a central station, with some of the circuits 25 miles in length.[28] This system lighted twenty-one square miles, and after disparagement and doubt during installation, was widely praised, drawing attention as far away as Honolulu.[29] A city official from Grand Rapids, Michigan, who leaned toward gas illumination inspected Detroit's system and was so agreeably disappointed that he became a convert to tower lighting. A journalist declared, "The press of the country has uniformly conceded the city to be the best-lighted of any in the world. All its streets, yards, alleys, backyards and grounds are illuminated as effectually as by the full moon at the zenith. The blending of light from the mass of towers serves to prevent dense shadows."[30] A Detroit tower took up no more space than a lamppost until it was 14 feet above the ground. Above that point, one could easily see through its lattice of metal elements. Standing on one leg, each was stabilized by guy wires and survived violent storms that uprooted nearby trees. The tall, thin towers did not dominate their surroundings. They lighted the city from half an hour after sunset until one hour before sunrise (see figure 4.3).

When Twain passed through Detroit on a lecture tour during the winter of 1884, he was quite taken with this form of lighting. He told a local journalist, "Your city is beautifully lighted by those electric towers. It is the handsomest appearing city at night that I have seen."[31] Twain also wrote in his journal about "the new light—there was nothing like it before. … & for the first time saw a city where the night was as beautiful as the day; saw for the first time in place of sallow twilight" from gas lights whose fuel had been "bought at three dollars a thousand feet" the far brighter illumination of arc light "clusters of coruscating electric suns" that on their high towers seemed to be "floating in the sky without visible support, & casting a mellow radiance upon the snow covered spires & domes and roofs & far stretching thoroughfares." The tower lights "gave to the spectacle that airy daintiness & delicacy of a picture & reminded one of the airy unreal cities caught in the glimpses of a dream."[32] The system Twain saw had not yet acquired the nickname "moonlight towers," and because he compared it to the dimmer gas lighting, the clusters of arc lights seemed like "suns." Tower lighting revealed the entire cityscape,

4.3 Woodward Avenue at Campus Martius, Detroit, Tower Arc Lights
Source: Library of Congress, Washington, DC

unlike conventional streetlights that emphasized only major streets, and Detroit acquired a dreamlike quality without being defamiliarized.

When the gas companies mounted a campaign disparaging Detroit's tower system, the public reaction was swift. A petition was signed by a large number of businesspeople expressing their satisfaction with the new system. A group of nineteen physicians declared, "We, the undersigned, practicing physicians in Detroit, having frequent occasions to visit all parts of the city during the nighttime, in all kinds of weather, have found the city well lighted by the tower system of electric light. We find the streets, yards, alleys, and public spaces lighted as if by moonlight, and have no hesitation in saying that Detroit is by far the best-lighted city that we know of. It is comparably better lighted than ever before by gas."[33] The citizens also appreciated that all the electric lights came on together, in contrast to the time-consuming lighting of gas lamps that started in the late afternoon and continued until 9:00 at night. In 1884, satisfaction with the tower system seemed almost universal. The *Detroit Journal* published a poem that read in part:

> Moonshiny, shimmering brightness,
> No planet hung on high,
> Gives to the world such dayness
> When nightness drapes the sky.
> Old Venus with her satellites
> Shrinks shyly out of sight
> When thou art open for biz,
> Aristocratic light![34]

In 1884, it briefly seemed that the tower system might replace all others. Passengers on steamers arriving in Detroit marveled at the illumination, and the system was favorably discussed in the *New York Times*.[35] On New Year's Eve, 1886, Chicago inaugurated a tower with 20 arc lights "a little over 300 feet above the ground" atop the Board of Trade Building. The city boasted it was "the highest and most powerful group of electric lights in the world."[36] Later that year, the National Electric Light Association met in Detroit, including experts from 147 utilities from all over the United States. The Brush Electric Power Company of

Detroit hosted the event, and the program on the first night featured a tour of its tower system. By this time, it boasted 576 arc lights, most in clusters of 4 on towers 150 feet high. The utility managers surveyed both the business district and suburbs to get a complete view of the system, which was served from one central plant.[37] From 8:00 until 10:00 p.m., a procession of eleven carriages drawn by white horses "wended its way through avenues, streets and alleys" of Detroit. "Although a general feeling of skepticism pervaded the minds of" the utility executives "as to the merits of tower lighting, it was almost entirely dispelled by a thorough examination of the actual illuminating power of the lamps."[38] The most convincing part of the trip passed through suburban streets.

> There are places in the city where a double row of beautiful forest monarchs reach out their joyous arms and kiss above the roadway below; and this too for half a block at a stretch. At the entrance to one of these high-arched primeval temples we would suddenly come pat upon an arc light on a telegraph pole, its daring brigand-like flicker and stare nearly blinding everyone, while everywhere apart from these uncanny spots, the diffused rays of a portion of the 576 tower lights, either too high above us to be unpleasant, or too far away to be dazzling, cast in every direction a gentle, soft, moonlight sort of illumination which was a complete and glorious surprise to all.[39]

After thorough inspection, most found Detroit's lighting more than satisfactory. A journalist from the *Electrician and Electrical Engineer* declared, "We have never experienced so general a change in opinion, caused by actual observation, as there was upon this subject. A large number of the visitors were strongly prejudiced against the tower system, but after thorough examination, became its earnest supporters."[40] On the second night of the conference, the lights could be seen to advantage again from the deck of a steamer during a four-hour excursion.[41] After that time, no electrical expert could remain ignorant of tower lighting, which continued to be adopted for the next five years. The city councils of Allegheny and Omaha, for example, sent representatives to ascertain whether such towers might be suitable for them, and both adopted the system.[42]

Yet for a combination of political and technological reasons, Detroit's tower system did not prevail. The city decided in 1890 to open competition between the Brush Company and another firm, which won the contract and erected a second system of towers and built another power plant. Brush was left with only private customers. Soon both companies were losing money and merged, but they had enormous overcapacity. Moreover, the public decried the merger as a monopoly and successfully agitated for a municipal utility, which gradually replaced the moonlight towers until Detroit was lighted like other major cities.

Tower lighting was not simply the casualty of political debate over public versus private power.[43] Detroit still used some of the towers as late as 1910, but it gradually replaced them. Several factors explain their disappearance. The tower system was labor intensive, though less so than gas lighting. A worker had to ascend each tower to replace the carbon rods and otherwise service the arc lights. They used a small elevator powered not by an electric motor but rather their own muscle power. This task was made somewhat easier because counterweights largely balanced the worker's weight, but it was still cumbersome and time consuming. One man could service only fourteen towers a day.[44] Detroit also abandoned its towers in part because trees blocked some of the light and occasional fog from the river reflected the illumination into the upper atmosphere. A few of the first towers also fell down in high winds or due to broken guy wires.[45] Tower light became less effective as Detroit built taller buildings. Furthermore, if towers with arc lights seemed to be marvelous in 1882, they appeared dimmer as other forms of lighting improved. Business interests wanted more powerful, focused illumination, and the towers gradually came down.

The installation of tower lighting caused conflict in Akron, Ohio, where two 207-foot towers were given a six-month trial. A member of the city council declared the towers made 325 of its 800 gaslights superfluous. He particularly liked the fact that "it lights up the whole area—alleys, back streets, vacant lots, etc.—that with any other system are not well-lighted, and must be the hiding places of those with evil intent."[46] He concluded that with multiple towers, there were fewer problems of shadows from foliage, and recommended the system for widespread adoption. In contrast, an author in the *Gas Light Journal* judged that light from the Akron towers was of practical value for a few blocks on a main street, but of little use on cross streets, where they lighted only "the tops of houses." The shadows were so black, he claimed, that a "footpad could lurk unmolested in any doorway," and it was impossible to see the faces of people walking away from a tower light.[47]

Both sides exaggerated; neither the virtues nor defects of tower lighting were as pronounced as claimed. A Chicago reporter interviewed many Akron citizens and found them divided on their new lighting system. Neither the city's hills nor its many trees favored towers over conventional streetlights. And yet some liked the new system, although many were doubtful. "By nearly all of the citizens it is regarded as experimental and not a sure thing." Members of the city council complained that the Brush Company had promised that the light would penetrate to a greater distance, and several businesspeople said some streets were too dark for a buggy driver to see pedestrians. The light was sufficient when unobstructed, but when blocked by buildings the darkness seemed all the more intense by contrast. A retired judge declared, "The time for lighting cities by electricity has not yet arrived." The mayor of Akron, however, firmly believed that when a full system was installed, rather than just two towers, "the light will be a success." He added, "One reason why the light has not been more satisfactory is that expectations had been placed too high. I live about a third of a mile from the mast; the light is shut off from me by a church and the heavy foliage of the trees." Nevertheless, he continued, "in the rear of my house, on a dark, cloudy night, I can see objects very distinctly. The light shows much better on a dark night than at any other time." He therefore was "of the opinion that six more masts with lights equal to those in use, will thoroughly light the city." A majority of local businesspeople and the president of the local bank disagreed. The bank president claimed that the adoption of electricity was the result of "a quarrel in which the gas company and the City Council figured."[48]

Despite controversy in Detroit and Akron, many cities bought tower systems. Fargo, North Dakota, which had few hills, installed a single 200-foot tower with 20,000 candlepower.[49] Omaha found the tower system "very satisfactory," not least because the city was more visible during a storm. The *Omaha Daily Bee* noted, "Night before last those who had occasion to be out during the heavy thunder storm had a peculiar sensation in noticing the rain pouring down" through the light "as if the moon could shine and the rain pour down at the same time."[50] San Jose, the first adopter in California, erected a central tower 237-feet high, with six arc lights that together had 24,000 candlepower. A reflector prevented light loss into the upper atmosphere and focused it on the city below. "The effect produced is very much like bright moonlight, the lights being so high that none of the direct rays reach the eye. For a distance of half a mile from the tower, in all directions, the light is brighter than would be produced by an ordinary gas-lamp every 75 feet."[51] A single tower made one-third of the gas lamps superfluous, and with four towers no gas would be required. Los Angeles was immediately inspired to imitation, urged on by the *Los Angeles Times*. In October 1881, the system was demonstrated by mounting arc lights on a water tower, and by the end of the year the city had erected 7 masts with 3 powerful arc lights apiece.[52] Each tower was expected to illuminate a radius of half a mile.[53] As this example suggests, rather than put 24,000 candlepower in one location, utilities began to place between 2 and 4 2,000-candlepower arc lights on each tower. The system was further refined in later years by varying the height of installations for different parts of town. Experience showed that high towers were poorly suited to city centers due to shadows cast by the taller buildings. The Jenney company found that "the best results can be obtained by placing lights on towers 125 feet in height in the outskirts where the houses are scattered" while using more conventional installations "at the street intersections in the business part of the city where the buildings are high and close together."[54] Davenport installed such a hybrid system in 1886. It contained "99 lights of 2,000 candle-power each, of which 40 are placed on 8 towers 125 feet high, 52 upon mast-arms 30 feet high, 5 over the intersection of streets, and 2 upon poles 40 and 50 feet high."[55] Similarly, in 1887, San Diego had 120 arc lights, using 10 towers, and poles at intersections, with some lights leased to businesses as well.[56]

Major ports early embraced electric tower lighting. As *Scientific American* observed in 1882, "The convenience and economy of electric illumination for harbors and water fronts, particularly when it is desirable to handle freight by night as well as by day, have been amply demonstrated in this country and in Europe." The Royal Albert Docks in London installed 27 arc lights on 80-foot towers in 1880. This enabled even the largest steamships to dock at night, and passengers could disembark and go through customs immediately instead of waiting until morning.[57] Liverpool also demonstrated "the economy of the electric light for docks and shipping, and the very satisfactory working of lights raised high in the air."[58]

The regular shipments of cotton from New Orleans to Liverpool drew the southern port's attention to this improvement, which it adopted. The site was ideal. "The crescent shape of the river front at New Orleans, the massing of the shipping business along a comparatively short reach of shore, and the broad open space along the levee to be illuminated conspire to make the elevated electric light especially serviceable and appropriate."[59] One hundred Brush arc lights of 2,000 candlepower were installed on the waterfront in 1882.[60] When Twain visited shortly afterward, he declared it "the best lighted city in the Union, electrically speaking. The New Orleans electric lights were more numerous than those of New York, and very much better. One had this modified noonday not only in Canal and some neighboring chief streets, but all along a stretch of five miles of river frontage." In fact, he remarked, "the finest thing we saw on our whole Mississippi trip, we saw as we approached New Orleans in the steam-tug. This was the curving frontage of the crescent city lit up with the white glare of five miles of electric lights. It was a wonderful sight, and very beautiful"[61] (see figure 4.4).

Pleased with tower lighting, New Orleans contracted with the Jenney Electric Company to illuminate the 1884 Cotton Centennial Exposition. One powerful electric tower of 100,000 candlepower called attention to the site from all over the city as well as 5 additional 125-foot towers that together had 36,000 candlepower, plus 125 Jenney arc lights elsewhere on the grounds.[62] So lighted, the fair remained open until 9:00 p.m. In the mid-1880s, New Orleans itself had 30 Jenney lights on towers in its public markets, 18 in public squares, and 771 mounted on poles for street lighting, together illuminating 30 square miles. Similarly, Pittsburgh first erected 22 arc lights on its wharves and soon after added them downtown.[63] Tower lighting made fewer inroads in the East, although in

4.4 New Orleans Levee at Night, 1883
Source: Library of Congress, Washington, DC

upstate New York, Utica adopted the Jenney system and Batavia erected 48 powerful arc lights, illuminating "a territory of one and half miles by two and one-half miles."[64] These systems provided service to all citizens on an equal basis. Putting arc lights high above the houses produced a general illumination that would seem weak to modern eyes, but appeared marvelous to a generation that had only known gaslight, whether in New Orleans, Pittsburgh, or Utica.

A town selecting tower lighting made both a technical and aesthetic decision. Moonlight had a high cultural value in the nineteenth century. As Matthew Beaumont notes in a discussion of Keats, "For the Romantics, night was a privileged time for apprehending nature, including human nature, in its least alienated, least mediated forms."[65] European composers wrote nocturnes—lyrical pieces inspired by the night—none more famous than Ludwig von Beethoven's *Moonlight Sonata*. On both sides of the Atlantic, the nocturne genre of painting depicted dreamlike night landscapes, often under a full moon. Nocturnes were painted by many leading artists, including James Abbott McNeill Whistler, George Inness, Albert Pinkham Ryder, and Charles Burchfield.[66] Walking by night had also become a romantic tradition. Henry David Thoreau emphasized the pleasing transformation of the landscape under a full moon. Night walking revived a sense of wonder and mystery.[67] Thoreau told readers of the *Atlantic Magazine*, "Chancing to take a memorable walk by moonlight some years ago, I resolved to take more such walks, and make acquaintance with another side of Nature." He thought a night landscape less profane as well as "more variegated and picturesque than by day. The smallest recesses in the rocks are dim and cavernous; the ferns in the wood appear of tropical size. The sweet fern and indigo in overgrown wood-paths wet you with dew up to your middle. The leaves of the shrub-oak are shining as if a liquid were flowing over them. The pools seen through the trees are as full of light as the sky."[68] In semidarkness his senses sharpened, and he noted sounds and fragrances missed during the day.

Nathaniel Hawthorne also liked to walk at night, and explained in a famous passage from *The Scarlet Letter* how the rays of the moon transformed and spiritualized everyday objects:

> Moonlight, in a familiar room, falling so white upon the carpet, and showcasing all its figures so distinctly,—making every object so minutely visible, yet so unlike a morning or noontide visibility,—is a medium the most suitable for a romance-writer. … There is the little domestic scenery of the well-known apartment; the chairs with each its separate individuality; the center-table, sustaining a work-basket, a volume or two, and an extinguished lamp; the sofa; the picture on the wall,—all these details, so completely seen, are so spiritualized by the unusual light, that they seem to lose their actual substance, and become things of the intellect. Nothing is too small or too trifling to undergo this change, and acquire dignity thereby. A child's shoe; the doll, seated in her little wicker carriage; the hobby-horse,—whatever, in a word, has been used or played with, during the day, is now invested with a quality of strangeness and remoteness, though still almost as vividly present as by daylight. Thus, therefore, the floor of our familiar room has become a neutral territory, somewhere between the real world and fairy-land, where the Actual and Imaginary may meet, and each imbue itself with the nature of the other.[69]

For millennia, human beings had witnessed moonlight transform everyday objects into something slightly strange and alluring. A city street under the alchemy of moonlight also became a site where the "Actual and Imaginary may meet." Twain sensed this in his description of Detroit under its new tower lights that had "the daintiness & delicacy of a picture."

Adoption of tower lighting was also a political decision. Twain's birthplace, Hannibal, Missouri, installed a Jenney tower system of 98 arc lights in 1883. The city operated the plant at a cost of $6,282.96 a year, two-thirds of which was recovered through rental of forty-six of the lights for commercial purposes.[70] Bay City, Michigan, also owned its arc lights, putting up towers for general illumination and paying for the whole system, plus a tidy profit, through the annual rental of eighty-seven lights at $100 each. Its city council declared, "We would not abandon our present system on any consideration. Even those who most bitterly opposed it are now entirely reconciled and satisfied."[71] Huntington, Indiana, owned its tower lights, and it too had saved money by doing so. As the mayor of Hannibal proclaimed, "All profits accrue to the people, where they rightly belong. You always have the matter under immediate control, and can so conduct it as to be of the greatest good to the greatest number."[72] Or as the city council of Flint, Michigan, concluded, "It may be justly called the poor man's light, for, by reason of its penetrating and far reaching rays, the suburbs of the city will be equally well lighted with the more central portions, and instead of the feeble flicker of the gasoline lamps, a clear and brilliant light will penetrate the most distant residential parts of the city."[73]

In these smaller cities, almost no house itself had electric lighting and tower lights wrought a fundamental change. The rhetoric of equality and democracy was common among city-owned utilities, whose goal was not merely to light individual streets but also the entire town. Hannibal wanted "an all-night light, of unequalled brilliancy." Such a system gave "the greatest satisfaction to the people, as there is not a dark street in the city." In a stellar example of US hyperbole, Hannibal's mayor asserted, "We claim the best-lighted city in the world."[74] This was a far-different ambition from that of the private utilities, which preferred to divide every commercial street into distinct areas—street, sidewalk, storefront, display window, electric sign, public buildings—each to be lighted for a price. They wanted competition between customers, not a nocturnal scene that closely resembled the city seen by moonlight. Where some saw poetry in the moonlit city, others, notably businesspeople, wanted to buy and sell focused brilliance.

When tower lighting fell out of favor, it was less a question of technology than of politics and culture. Arc lights continually improved, and after 1910 tungsten incandescent lights (requiring less frequent service) could have been substituted. Nor was it simply that hills or taller buildings blocked tower lights and created strong contrasts and dark shadows. An official in Utica concluded that "the towers are used mainly in the outskirts and thinly settled districts. There they are a perfect success. In the heart of the city they are a failure."[75] In Cleveland, the same objections were raised, as narrow streets, tall buildings, and trees blocked the light.[76] Yet a hybrid system like that used in Davenport solved this problem, with arc lights on conventional poles at major intersections and tower lighting for other areas. Some objected to the towers as eyesores, notably the English electrician William Preece, who complained of the "unsightly posts."[77] But few Americans seemed to mind the towers themselves. What undermined moonlight towers was the demand for more powerful illumination in business districts, including that for advertising. Later, the few tower lights that remained were replaced because the spread of automobiles made lighting focused on highways necessary for public safety.

The emerging automotive center of Detroit sold some of its towers to Austin, Texas, where a few are still operating as one of the last remnants of this early system.[78] If some, like Twain, found artificial moonlight poetic, others wanted rows of electric streetlights. Apparently no city went from tower lighting back to gas—proof that it was thought better than what had seemed adequate before. By 1888, Wabash had abandoned towers for incandescent streetlights—an unusual choice at that time.[79] Whatever the choice, the public wanted ever more illumination, as electricity shifted from being novel to seeming indispensable. Tower lights were cheaper than streetlights, but many citizens wanted more than moonlight.

The nocturnal landscape of tower lighting replicated natural light sources. Rays of light had always fallen from the heavens, and the full moon and stars on a clear night provided a plausible standard for night illumination. Rows of lights closer to the ground produced a fundamentally different sense of space, in which the visible was primarily at street level, while the higher floors of buildings were darker, and the sky darker still. In some towns, strands of lights hung over the street, creating a sense of enclosure, as if there were a ceiling over the street. This use of luminous arches and festoons was particularly popular at amusement parks and during carnivals. Tower lighting supplied something quite different: the equivalent of the best natural level of light. It was designed to present a town's features without discrimination or distortion, as an entirety. The urban landscape under tower lights closely resembled the city by day, while conventional streetlights left most of the visual field in darkness or shadows. They showed the pavement, but the higher floors of buildings and back streets were scarcely visible. Tower lighting was at once radical in its wide diffusion and conservative in its goal to keep the entire city visible, rather than singling out some parts over others.

Gas lighting had accustomed the public to a low-level partial view of the streets, which only implied the remainder of the largely invisible night city. The arc light's more intense illumination could either be used to simulate moonlight or intensify lighting at ground level. Americans chose to intensify patterns familiar from gas lighting and create a night cityscape dramatically different from that of the day. This new landscape was increasingly shaped by private investment, and only a few remnants of tower lighting survived.

Until the early 1890s, however, tower lighting was widely adopted and seemed to be a good solution to the problem of lighting public space (see figure 4.5). The superintendent of the Chicago Electrical Department, John F. Barrett, compared electrical systems to determine which to use for "the Chicago River, wharves, bridges, and slips." He considered existing gas lighting inadequate and concluded that "for points where powerful concentration of the light is required, as at the bridges, along the river, viaducts and parks, the 2,000 candle-power arc light is far preferable." Where less intense but even light was needed, the incandescent system seemed "the more advantageous." Yet the cheapest incandescent system then cost 25 percent more than arc lighting. Barrett concluded, "The arc lights should be used in the business portion of the city, and the incandescent in that portion of the city where foliage, shrubbery, etc., is located, and where more equal distribution of the light is required."[80]

Moonlight towers spread an even illumination in the transitional cityscape of the 1880s, when almost no houses were yet electrified, before the development of giant advertising signs or powerful special effects that drew the eye to particular places. No areas were cast into darkness and obliterated as unimportant blanks. The moonlight towers arose before electricity had become a central part of a dynamic night environment of new activities. They did not redefine the city as a site of after-hours dance halls, theaters, roller-skating rinks, amusement parks, and spectacles. Tower lighting was without special effects. It did not accentuate commerce, and did not heighten the division between dark residences and a brilliantly lighted vortex of pleasures.

4.5 Tower Arc Light, Los Angeles, ca. 1890
Source: USC Digital Library, California Historical Society Collection at the University of Southern California, Los Angeles

A transitional technology, tower arc lighting was brighter than gas lighting, which because of it never gained much of a foothold in many western cities. Tower lighting accustomed communities not just to a brighter night but also to a mild transformation. Under artificial moonlight, a town did not assume an entirely new appearance, nor did it become as sepia toned as it did under gaslight. Rather, its colors faded somewhat toward a black-and-white version of itself that seemed calm and even dreamlike. Tower illumination balanced between nature and culture, and did not surpass moonlight.

This aesthetic increasingly competed with two other possibilities examined in the following two chapters—one embodied in the intense visual displays of the great world's fairs, and the other in the commercial centers of the largest cities, most famously New York's Great White Way. This three-way competition raised many questions. Should illumination be democratically and equally dispersed, or focused in central zones? Should the night city be designed to appear individualistic or cooperative? Should it be energetic or calm, dynamic or dreamlike, technologically beautiful or technologically sublime? By 1915, it was clear that US cities would adopt a hierarchy of illumination that concentrated on the downtown, crowned by corporate advertising and skyscrapers. In the second rank came monuments and public buildings. Next came local businesses with their signs and lighted windows, followed by the fashionable residential areas. Last came blighted areas, rendered dim or invisible, cast into a darkness that was all the more absolute in contrast to the soaring illuminated towers, powerful streetlights, and flashing signs of central districts. This hierarchical system intensified illumination at a few sites, and erased poverty and visual blight. It aggressively transformed appearances, creating a heterotopian space where two or more places could be experienced. This place making divided the city into zones. It created a hierarchy of attention within a competitive system. By the 1890s, the moonlight towers were coming down, leaving as their legacy the expectation that every neighborhood deserved some lighting. Yet increasingly, the illumination of US cities visualized competition and the disruptive aesthetics appropriate for a capitalist society.

The Italian futurist Filippo Tommaso Marinetti made explicit "the linkage between the advance of modernity and the retreat of the moon."[81] He described how "a cry went up … 'Let's murder the moonlight!'" In response, "gigantic wheels were raised, and turbines transformed the rushing waters into magnetic pulses that rushed up wires, up high poles, up to shining, humming globes. So it was that three hundred electric moons cancelled with their rays of blinding mineral whiteness the ancient green queen of loves."[82] For Marinetti, the moon was an antiquated emblem of superstition and tradition, to be jettisoned in favor of the bracing, night-destroying electric light. This was more than a rupture with the natural world. A hierarchical form of electrification also increased the distance between and relative visibility of the social classes.

A century later, the Civil Twilight Design Collective won an award for a "lunar-resonant streetlight system," which brightens and dims in response to the ambient level of lighting, with bright moonlight being the standard level sought. The collective argued for "utilizing available moonlight, rather than overwhelming it" as a way to save energy, reduce costs, curb light pollution, and recover the experience of nighttime as "one of the most fundamental and beautiful cycles of nature."[83] Using interactive computer technologies, the system reinstated natural moonlight as the norm. Perhaps one day US cities will adopt a lunar-resonant streetlight system, but first they will need to tone down the brilliant night landscape they created and naturalized between 1880 and 1915.

5

Spectacles and Expositions

Even as tower lighting spread in the Midwest, South, and far West, subtler and more powerful forms of lighting were being displayed in spectacles around the country. The public appetite for brilliant illuminations was already evident during the gas era, including celebrations of the first transatlantic cable, the end of the Civil War, and every Fourth of July. The political parties had adopted illuminations as part of election campaigns, including transparencies, marchers carrying torches, gas jets, profusions of Chinese lanterns, and fireworks.[1] In the 1870s, there were illuminations to celebrate centennials of the Declaration of Independence and other events of the American Revolution. In 1881, no electricity was utilized at the Yorktown Battle site, where President Chester A. Arthur and foreign dignitaries observed pyrotechnics bursting above a fleet of ships "illuminated with thousands of lanterns of every color, forming a scene of great beauty" until rain and gale-force winds necessitated "a sudden termination."[2] The Yorktown Centennial was perhaps the last important US illumination without electrical effects. In Washington itself, the anniversary featured Brush arc lights placed on the capitol and treasury buildings to light up Pennsylvania Avenue.[3]

Incandescent lighting enabled mobile illuminations, as Edison dramatically showed in the same year, when a procession of his employees marched through New York's business district (see figure 5.1). Each person wore a specially designed helmet with a light bulb on top so that a parade of lights moved down the street. To achieve this effect, the employees marched in hollow-square formation while pulling an electric dynamo on a wagon. Wires ran from the dynamo to each marcher and up

5.1 Edison Workers' Parade
Source: Hall of History, Schenectady Museum, Schenectady, NY

their sleeve into the helmet. Edison's stunt was widely reported and even copied in 1884, when Cheyenne, Wyoming, staged a similar procession. There, batteries in a horse-drawn wagon powered the lights of marchers in a procession that celebrated the selection of a new governor.[4]

Spectacular lighting was prominent in theaters. One of the leading innovators, Steele MacKaye, was an actor, playwright, theater owner, and inventor of scores of devices used in theaters. Edison personally advised him and supplied lighting for a remodeled New York playhouse.[5] Gas had been used primarily for footlights, but electric lighting could be mounted anywhere with little risk of fire, and the new installations provided a light closer to daylight. Subtle effects could be realized, including the slow intensification of the brightness of a scene, rapid alternation of colors, or sharply focused, moving shadows. By the 1890s, most theaters had adopted such effects and made lighting an integral part of rehearsals.[6]

Other public events used lighting to become more theatrical. P. T. Barnum's circus purchased arc lights from Brush in 1878, and they became a popular part of his show. When the circus came to Washington, DC, in April 1879, the nation's capital had no generating plant. Barnum brought one with him, and the lighting of the main tent was considered an outstanding feature of the performance.[7] Two years later President James Garfield's inauguration combined gas, fireworks, calcium lights, and arc lighting. The ball at the Smithsonian seemed "a Crystal Palace. The contrast between the whiteness of the electric lights in the rotunda and dome, and the yellowness of the thousands of gas burners elsewhere, produced a very fine effect."[8] Gas burners were installed especially for the event and taken down afterward, as the Smithsonian normally closed at night. Pennsylvania Avenue was the first Washington street to replace gas lamps with arc lights, but this was not done until 1886.[9] The city's nocturnal landscape was then heterogeneous, including gas, arc lighting, and incandescent bulbs, each affiliated with a different manufacturer.[10]

Many cities had several forms of illumination, including at a minimum an older gas lighting system, arc streetlights, and incandescent lights in residences and commercial buildings. Saint Louis used all these in its annual event, the Veiled Prophet. Started in 1878 by local businesspeople and prominent families, the event grew into a six-week spectacle with increasingly elaborate illuminations. In 1882, it featured twenty thousand individual globes of various colors illuminated by gas. In the first years its parade floats were "kept in a blaze of light by torch-bearers marching on both sides," until electricity provided a more brilliant alternative.[11] By 1883, the parade had evolved into a line of floats, brass bands, and columns of marchers, led by the Veiled Prophet (see figure 5.2). A German newspaper reporter that year joined a crowd of a hundred thousand. "The streets have been wonderfully illuminated. For miles, arches with gaslights have been hung under red or white cupolas all along the route taken by the parade on the first day of the fair. Interspersed with the arches are gas lanterns fastened to poles. These endlessly long streets and avenues so gloriously lighted are really a sight to behold." In addition, there was "extensive use of electric lighting. Individual business establishments have hung dozens of arc lamps from extensions in their store fronts."[12]

The Veiled Prophet took inspiration from New Orleans's Mardi Gras, but was held in the early fall so as not to compete with it. As in New Orleans, a secret organization was in charge. The businesspeople who invented and organized it wanted to reassert the importance of Saint Louis, whose commerce had been strangled during the Civil War, which halted most trade on the Mississippi River. It had lost markets to Chicago. The Veiled Prophet was to rally and unite the city, and advertise it to the rest of the country. By 1893, the Veiled Prophet parade used seventy-five thousand lights, more than half of them electric, and it attracted travelers on their way to or from the Chicago Columbian Exposition.[13] Saint Louis advertised in newspapers as far away as Canada and Mexico, and arranged for reduced railroad fares for those attending the event, which lasted from September 4 until October 21. There were "Four Concerts Daily by Sousa's Grand Concert Band" and nightly illuminations, but the festivities focused on nine weekend dates when the largest crowds could come. Saint Louis used every available lighting technology, and erected "electric fountains, electric panoramas, electric revolving stars, electric flags, electric arches," and "electric portraits," intermingled with "myriads of gas jets, gas transparencies, gas arches, gas combinations, and gas clusters."[14] Visitors were promised "Magnificent Street illuminations which excel in extent and gorgeousness any illumination ever attempted in any city of the West." The climax was "a panorama depicting the discovery of America, its gradual settling up, and the

5.2 Veiled Prophet Parade
Source: Cox, *St. Louis through a Camera*

final triumph of prosperity and civilization." The organizers erected "a massive pedestal 125 feet high" topped by a globe showing the outlines of North and South America. When the Veiled Prophet's float neared the column, a bright star appeared where Christopher Columbus first landed at San Salvador and "simultaneously the date 1492" also "burst out in relief." After this, the "march of discovery and civilization" was "shown by electric lights, until finally the entire continent" was "outlined by a series of hundreds of electric lights." The brightest star then flashed on above the globe, accompanied by the date 1892.[15] As many as seventy-five to a hundred thousand people a day saw the electrical panoramas, concerts, fireworks, and parades.[16]

The annual event also had latent functions that organizers in later years were perhaps not entirely aware of. As Thomas Spencer notes, a former Confederate cavalry soldier, Charles Slayback, invented the Veiled Prophet after the great 1877 railway strike, when Saint Louis witnessed the first general strike in the United States. Building on previous organizing by the Knights of Labor, the strike had united workers from many industries. For one week in July the strikers controlled Saint Louis, until three thousand federal troops and five thousand special police dislodged them, killing eighteen strikers and wounding many more. A year later, the first Veiled Prophet was the police chief who had suppressed the strike.[17] The organizers were the cultural and business elite who took for granted their right to lead. They were concerned that workers and immigrants did not share their sense of history and civic values. The parades were not mere entertainment but celebrated those whom they regarded as great individuals, and inculcated advice on social and moral issues.

The Veiled Prophet also offered an alternative to carousing and radical politics. Its parades, fireworks, exhibits, and displays took back control of the streets and sought to awe and instruct the public. Its floats were didactic. For example, in the event's eighth year, the floats presented the history of the New World in twenty episodes, beginning with the "goddess of liberty," the Norse discovery of America, Columbus's voyage, the discovery of the Mississippi River, Pocahontas and John Smith, Henry Hudson's voyages, the Pilgrims at Plymouth Rock, George Washington crossing the Delaware, and on down to the final float, where the Veiled Prophet sat "on a massive throne of gold."[18] The organizers convinced President Grover Cleveland to attend in 1887, drawing enormous crowds to a parade that featured Old Testament themes from Adam and Eve to Daniel in the Lion's Den. The procession was illuminated by colored electric lights and "hundreds of thousands of gas jets especially erected along the streets."[19]

The Veiled Prophet did meet opposition. Some youths with peashooters annually showered the floats with dried peas. In later years, "working-class boys" would "attempt to knock the metal pole loose that conducted the electricity" from the trolley lines to the floats. If they succeeded, the entire parade halted, and "lights would flash off and on" until an attendant reattached the pole to the line. Protests increased after police killed three strikers and wounded fourteen in the Saint Louis streetcar strike of 1900.[20] A labor publication sneered that the event ought to be called "the veiled profit," remarking that it was only "some paint, tinsel, and electric light."[21] The spectacle was a distraction; it did not resolve class conflict.

When parades avoided controversy by being less educational and more entertaining, organizers worried that their events lost their civic utility. Some cities began to use electrified floats devoid of patriotic content. Electric trolley cars were rapidly being adopted in the 1890s, and they easily could be turned into the floats in a night parade with brilliant lighting effects.[22] In 1893, Wilmington, North Carolina, was one of the first to do so. Milwaukee used streetcars for twenty parade floats in 1900. Illuminated by thousands of colored lights, they had no didactic purpose, but were presented "solely for their picturesque beauty and scope for fantastic display of form, color, and effulgent light." The floats included two giant swans, "the Goddess of Light on a sun bristling with golden rays," a replica of the battleship *Wisconsin*, and Jonah inside an electrified whale.[23] As such unscripted events proliferated, David Glassberg observes, "members of the educational and hereditary elite worried about a growing commercialization of municipal ceremonies and concomitant infusion of a carnival atmosphere."[24] This was especially worrisome because, it seemed, the flood of immigrants knew little about the history or values expressed in events such as the Veiled Prophet. Because spectators enjoyed fireworks, parades, and entertainment more than speeches, comprehensive didactic displays seemed necessary, notably at world's fairs.

Modern world's fairs began in 1851 with the London Crystal Palace Exposition, and their number and size increased until World War I. After 1881, spectacular electrical displays became a central part of these events. In Europe, national governments played a guiding role, but in the United States, local organizers had to conceive of a fair, forge a citywide coalition of government, business, and cultural institutions, gain support from the federal government, and convince other countries to participate. A US exposition needed support from the press, social elites, schools, universities, churches, ethnic associations, and worker organizations. Expositions also encouraged professionals to attend by hosting hundreds of congresses. (This inclusivity usually did not extend to blacks, however, and embraced women less than men.) The most numerous and enthusiastic visitors were middle- and upper-class families; expositions provided them with a vision of progress and glimpse of what future cities might look like.

Until 1880, world's fairs featured steam technology. Victorians understood the general principles and mechanics of steam engines, with their boilers, pistons, drive shafts, gears, belts, valves, pulleys, cutoffs, and levers. The transmission of steam power was a visible process that could be traced with the eye and hand. It had come to seem a part of common sense. Early fairs erected huge steam engines, which the public admired for their size and smooth operation. The apotheosis of this trend was the giant Corliss engine at the Philadelphia Exposition in 1876 that drove all the devices in Machinery Hall. The fair's central symbol, its enormous flywheel, moved just slowly enough so that the eye could follow its silent revolutions.[25]

The 1851 Crystal Palace Exposition had closed at dusk, but later fairs relied on gaslight for illumination and impressive displays. The Milan Exposition of 1881 was one of the last to rely primarily on gas jets, including two hundred thousand that highlighted architectural details on the facade of the cathedral, in combination with arc lighting. Overhead, arched strings of colored incandescent bulbs created the illusion of a ceiling.[26] Electricity quickly became dominant at US expositions, but gas persisted at European fairs. At the 1889 Paris Exposition, the Eiffel Tower was lighted by a combination of gas, arc, and incandescent lights, and in the 1900 Paris Exposition, Welsbach gaslights were used extensively.

But Americans regarded electricity as the cutting-edge technology of the time, and it had the same prestige that computers and cell phones enjoyed a century later. They presented the United States as the world leader in electricity, and used electric illumination as a universalizing symbol that pointed toward a utopian future.[27]

As the scale of fairs increased, they adopted comprehensive architectural plans. "Everything contrasted with the usual city: impeccable, grandiose buildings, constructed with virtuosity; perfect urban planning; beautiful, healthy, and well-tended flower beds; freshly swept passages; abundant urban furniture in the form of fountains, sculptures, and lampposts; luxuriant lighting." The fairground "was a sort of 1:1 scale model of an ideal city."[28] Lighting expanded their hours of operation and attracted large night crowds. Expositions educated visitors in the spectacular possibilities of electricity. They specialized in dramatic illuminations and special effects that accustomed visitors to transformations of space, and awed crowds every evening with sparkling landscapes.

Most European expositions were held in capitals, notably London, Vienna, and Paris, which were brightly lighted already, but Americans held expositions in regional centers. The crowds in Louisville, New Orleans, Atlanta, Nashville, Omaha, and Buffalo were less accustomed to advanced lighting, and the electrical displays therefore seemed all the more dazzling. Crowds were often larger at night, expressing their enthusiasm for the electrical sublime. Even the experienced and usually critical Saint Louis newspaper reporter young Theodore Dreiser was overwhelmed. Viewing the electrified Chicago Columbian Exposition from a launch in the central lagoon, he felt transported into an Elysian realm: "A feeling of the true dreamlike quality of it all came to me, at first only as a sense of intense elevation. ... Then followed an abiding wonder."[29]

In contrast to the steam technologies that dominated early fairs, electricity was instantaneous and invisible, working according to principles few could grasp. Even the Harvard-educated Henry Adams complained that after viewing electrical exhibits at many world's fairs, he still did not comprehend the new force.[30] Every fair after 1881 devoted a central building to electricity, and these were among the largest and most popular. The telegraph, telephone, phonograph, electric light, and loudspeaker were at first displayed individually, as was the case at the 1881

Exposition Internationale d'Électricité in Paris.[31] When it opened, the exposition had 277 arc lamps, 44 arc incandescent lamps, and 1,500 incandescent lamps. The number was increased to 2,500 lamps in the subsequent months.[32] It was impressive and even overwhelming to most visitors, as electric light was so new. Yet William Hammer, who erected the Edison exhibit, concluded that "no manager of any future exhibition is likely to repeat that terrific *mélange* of lights that flooded the interior of the Palais de l'Industrie with great brilliancy, but with an impracticable and impossible means of comparing and judging the relative merits of different systems."[33]

France had 55 percent of the exhibits, yet there was also significant participation from Germany, Great Britain, Belgium, and the United States. Visitors traveled on an electric train, built by Siemens, which took them five hundred meters from the entrance to the main exhibit hall.[34] Its ten thousand square meters were organized by nationality and biography. Edison and Alexander Graham Bell were prominent on the main floor. Two rooms dedicated to Edison featured not just the light bulb but also his entire system, with its generators, underground cables, wiring, fuses, sockets, and the light bulb itself. Bell displayed not just a working telephone but a telephone exchange too. Visitors left such exhibits with intimations of what an electrified world might be like. Electric lighting was an ideal element of display, at once flexible, refined, abstract, and modern. Spectacular lighting did not detract from other exhibits but rather connected diverse elements into one stunning design. Many fairs made illuminated towers their central symbols, including the Eiffel Tower, Buffalo's Electric Tower, and San Francisco's Tower of Jewels. Electricity became more than just the focus for one building; it provided a visible correlative for the ideology of progress.

Electrical corporations became deeply involved in overall fair planning. For the Louisville Southern States Exposition of 1883, Edison sent Luther Stieringer to work with the fair's architects. They created a comprehensive lighting plan with 7,000 16-candlepower light bulbs in the exhibition's 14-acre interior, offering visitors their first experience of a vast electrified interior.[35] Stieringer knew earlier illumination traditions, and drew inspiration from the ornamental use of gas and Chinese festive illuminations.[36] Louisville had twice as many lights as Paris two years before, and the 1884 International Electrical Exhibition in Philadelphia had even more candlepower. The "four competing electric light companies" hung arc and incandescent lights "from girders and arranged in rows along the thirteen arched rafters of the main building."[37] There were "5600 incandescent lights and 350 arc lights" to illuminate 1,500 exhibits. A reporter noted "the crude buildings hurriedly erected without any attempt at finish for a temporary purpose, were transformed into a temple of light, which at the first glimpse evoked expressions of delight from every beholder."[38] Another reporter remarked that in "New York so much interest has been taken in the affair that there is scarcely an electrician to be found in the city." In Philadelphia, in just one section, they saw "electric telegraphs, telephones, microphones, etc., fire and burglar alarms, annunciators, electric clocks and time telegraphs, electric registering and signal apparatus, applications of electricity to dentistry, to warfare, to mining and blasting, to spinning and weaving, to traps and snares, to pneumatic apparatus, to musical instruments, to writing and printing, to conjuring apparatus and to toys." The exhibit also featured an indoor fountain 30 feet in diameter, whose 15 jets periodically became the center of attention after extinguishing all other illuminations and training colored lights on the streams of water. Equally striking was a 30-foot column covered with 1,000 bulbs that "climbed" around it.[39] The largest Edison dynamo in 1884 impressed the crowd because it could supply 1,200 16-candlepower bulbs. That would seem small a decade later.[40]

The level of illumination increased further at the New Orleans Cotton States Exposition of 1884–1885. The *Boston Herald* thought the main building "one of the most remarkable ever erected in the United States," but the *American Architect and Building News* called it an enormous, unfinished barn, although the lighting was "nearly perfect as it can be."[41] Its 33 acres of interior space required 20,000 16-candlepower incandescent lights, which were particularly appreciated in the art galleries and enormous music hall. Outside, tower arc lighting lit up the grounds.[42]

In 1888, Cincinnati hosted the Centennial Exposition of the Ohio Valley and central states. The city itself relied largely on gas for illumination, which made the 100 powerful arc lamps outside the exhibition buildings a striking contrast.[43] Inside, visitors were stunned by arrays of incandescent lamps from most of the companies then producing them.[44]

Lighting was rapidly becoming a form of entertainment. Hammer, again Edison's exhibition expert, put faint light bulbs in the eyes of stuffed owls, carved animals, and Japanese paper fish, and was praised for producing "color effects with the skill of an aquarellist." The *Toledo Daily Bee* described "an electric rotating flower garden 14 feet in diameter and 12 feet high banked with rare exotic flowers, with colored incandescent lamps for petals that are constantly budding, blooming, fading, and dying, six times per minute, the entire garden rotating constantly." One reporter considered the illuminated Horticultural Building a "beautiful boudoir of electrical brilliancy." Another called the electrical displays "one long rainbow through the night" that outdid the Alhambra.[45] Some attendants in the entry hall wore "helmets surmounted by a powerful electric lamp" that blinked on whenever they stepped on a steel plate. The most impressive effect was an "electrical cascade" where 7,000 gallons of water a minute flowed "over a bed of electric fire, constantly changing color and scintillating." It threw a jet of water 100 feet high in the midst of a luxuriant garden scene embellished with hundreds of colored lights, while overhead twinkled "tiny stars, moons, and crescents."[46]

In contrast, at the Paris Exposition of 1889, more than half the indoor exhibits were closed after dark.[47] Instead, "a nightly show of illuminated fountains entranced crowds with a spectacle of falling rainbows, cascading jewels, and flaming liquids, while spotlights placed on the top of the Eiffel Tower swept the darkening sky as the lights of the city were being turned on."[48] There was "an exodus from Paris every night to the Exhibition." The *Pall Mall Gazette* declared that the "spectacle, even on ordinary nights, is one of unparalleled beauty. ... During the day, people go to the exhibition partly for instruction; at night they go solely to be entertained and to witness a brilliant spectacle." In fact, the price of admission was higher in the evening, even though many exhibits were closed. For this event, Hammer created a display 15 meters tall that contained 25,000 colored bulbs. It depicted the American and French flags, waving and flashing as the words "Paris" and "Edison" appeared, and streaks of lightning seemed to run up and down the sides. The fair as a whole outlined "the prominent architectural features of the building facades with light," an "amplification of effects previously obtained with gas jets."[49]

Hammer's colleague Stieringer, who began his career as a gas illumination expert, developed exposition lighting further in a series of US fairs between 1893 and 1901.[50] Like Hammer, Stieringer was a close Edison associate.[51] He was the chief consulting engineer for the Chicago Columbian Exposition, which attracted 27.5 million visitors, almost a third of the US population.[52] The exposition required five times more power for lighting than the Paris Exposition had just four years before.[53] Visitors from small towns saw more artificial light there in a single night than they had in their entire lives (see figure 5.3). Attendance peaked in the cool of the evening, when the noiseless electric launches in the lagoons were packed and crowds promenaded around the grounds. As one journalist described it, "Just as the first stars came out under a mistaken idea that it was their right to shine, the administration building put on its jewels and the crowd around the plaza saw a building beautiful as a fairy tale. Encircling the cornice was a band of lights, joined to which were strands sparking with electric gems which lighted the building from dome to cornice." At the top of the dome "was a cluster of brilliant arc lights," while inside it was "illuminated to a noonday degree." The grounds were thick with visitors, who "reveled in the feast of sight and … the music" provided by two bands. At each end of the Court of Honor, electric fountains shot forty-four thousand gallons of water a minute in kaleidoscopic patterns against the night sky. Beneath the fountains, spotlights fitted with colored filters permitted operators to create symphonies of color to the accompaniment of the band music. The electrical engineers thought these fountains in the Court of Honor "taught the public the possibilities" electricity offered for inexpensive improvements to city parks.[54] To see the entire panorama, many paid a quarter for an elevator ride to the roof of the Manufacturer's Building for a "view that was simply dazzling." It included not only the fairgrounds but also the city skyline in the distance, blast furnaces of South Chicago's steel mills, and distant steamships on Lake Michigan.[55]

Lighting inside the exposition buildings treated the long aisles as city streets, with arc lights on ornamental posts. These were "shielded" to provide "an opalescent glow rather than a fierce, sputtering spark," while individual exhibitors used incandescent bulbs.[56] Inside the Electrical Building were 450 arc lights shedding a hundred times more illumination

5.3 Columbian Exposition, 1893
Source: Danish National Library, Copenhagen

than had seemed so spectacular in Wabash just a decade before.[57] At the center of the building was a much larger and more complex "Tower of Light" than had been displayed in Philadelphia. It was covered with 10,000 incandescent lamps, surmounted by a "mammoth incandescent lamp built up of about 30,000 cut glass prisms. ... By means of a commutating device within the base of the tower, the lights were thrown on and off in a great variety of charming combinations."[58] It drew attention to an exhibit about light bulb production, but such educational displays took more time and effort to view than spectacular lighting.

Early every evening, crowds began to gather around the Court of Honor "eager to secure a good position from which to behold the illuminations." Yet the spectacle covered "so wide a range of territory, that it is no easy matter to obtain a position where all can be surveyed. The electric fountains and Administration Building in a blaze of glory are at the west end; the magnificent pyrotechnic display is eastward of the lake; the surface of the grand basin is covered with floats from which shoot up numberless fiery serpents; all along the roofs of the Agricultural and Liberal Arts Buildings are lines of flickering flambeaux."[59] Many, including Adams, preferred the panoramic view from the Ferris wheel.[60] This landscape of light charmed even the most critical observers.

The 1898 Trans-Mississippi Exposition in Omaha hired Stieringer and Henry Rustin to design an entirely new kind of illumination.[61] All but abandoning arc lighting, Stieringer and Rustin developed the first "comprehensive and complete decorative and ground illumination by incandescent lamps." They created subtle effects with 21,000 8- and 16-candlepower bulbs, and festooned the fair's neoclassical buildings with bulbs, supplemented by clusters of incandescent lamps on 309 posts. They hid wiring underground. By using so many small lights, "concentrated or intense light was carefully avoided" and dark shadows eliminated (see figure 5.4).[62] At dusk on opening day, a vast crowd filled the Omaha grounds.

> Just as the outlines of the faraway buildings began to grow indistinct in the deepening shadows, a single cluster of electric lamps on each side of the lagoon was lighted. Then another and another until the row of pillars that circles midway between

5.4 Grand Court, Omaha Exposition, 1898
Source: Omaha Public Library

the lagoon and the buildings was crowned with incandescent luster. Another turn of the switchboard and the circle immediately surrounding the lagoon added its radiance and flashed golden bars across the water. In another instant the full circuit was opened and every outline and pinnacle of the big buildings blazed with light. The effect was indescribable. … It was magnificent beyond comparison or comment and the immense crowd that had been waiting patiently for the moment gazed in dumb admiration. For a few seconds the vast court was as silent as though it was peopled with wax figures. The approbation of the people was vented in a volley of cheers and handclapping. On every side were heard the most extravagant expressions of admiration.[63]

Harper's Weekly thought the effects "superb," and far more subtle and beautiful than an illumination that its reporter had recently seen in Paris.[64] Stieringer received a gold medal and diploma from the satisfied exposition organizers.

The 1900 Paris Exposition only haphazardly adopted the techniques pioneered in Omaha. The *Revue de Paris* marveled that a "simple touch of the finger on a lever" could transform the Monumental Gateway with "the brilliance of three thousand incandescent lights, which, under uncut gems of colored glass, become the sparking soul of enormous jewels."[65] The banks of the Seine and its bridges were also illuminated, giving Parisians a foretaste of the electrified urban landscape that soon would become normal. Yet the Paris fair used far fewer lights than Omaha. It was only half the size of that in Chicago, and the electrical exhibit covered only a moderate area in one building. Paris selected the Welsbach gas mantle to light the grounds, and the only enclosed arc lamps exhibited came from the United States, where this technology was being widely adopted. The exposition was a popular success, but the trained eye saw that French engineering was falling behind.[66] International experts concluded that the most impressive technical exhibits came from Britain, Germany, and the United States.[67] The Paris Exposition had sent a committee to investigate the illumination in Omaha in 1898, where Stieringer explained his innovations to them. But the Paris Exposition proved to be "a distinct step backward," without a "uniform scheme of illumination. ... The lighting was a mixture of large and small incandescents, searchlights, projectors, display lighting of the spectacular order, acetylene, Nernst lamps, Welsbach burners, gas, and other illuminants—a conglomeration which was entitled to more credit as an exhibition of all known modern forms of lighting than as a comprehensive scheme of exposition illumination."[68]

In contrast, the 1901 Buffalo Pan-American Exposition had a coordinated lighting design, and electrification was the fair's central theme. It celebrated the new hydroelectric power stations at nearby Niagara Falls and improved systems of AC that transmitted electricity long distances. In Buffalo, Stieringer expanded on the Omaha model. He used no arc lights but rather an immense number of eight-candlepower bulbs that subdivided the light so thoroughly that it eliminated glare. At the center of the fairground stood the four-hundred-foot Electric Tower, covered with forty thousand bulbs. The surrounding Great Court had two hundred thousand more. The number of lights increased as one neared the Electric Tower, which was "the climax of the lighting scheme."[69] Stieringer's plan was based on the understanding that "to secure the best results it is essential in all artificial illumination to get uniform diffusion of light," which could not be achieved "by the arrangement of a few extremely brilliant centers about the space to be lighted," such as arc lights or tower lighting. "In fact, such an arrangement acts in itself to defeat the object to be attained," because most of the illumination will be "immediately about the sources while the lighting dims" as one moves away from each source "in obedience to the unvarying law of inverse squares." A far more minute subdivision of the light was achieved with "a lamp so small, as compared with those now in common use, that it gives but little light individually, but is capable of being so grouped, massed or distributed as to produce" many different effects, "without raising any point of space to a brilliancy disagreeable to the eye to rest upon." Stieringer used "the beauty which is inherent in the light itself" and avoided merely sensational displays.[70]

The Buffalo exposition also required a telephone exchange, fire alarm system, and heating as well as illumination. Wiring the central area around the Court of Fountains required "250 tons of insulated copper wire of all sizes," with more wiring inside the buildings.[71] *Scientific American* compared the work to that of a stage designer, dealing with a space the size of twelve football fields. The central basin was not merely a mirror reflecting the illuminations; it contained floating lights and ninety powerful searchlights. The entire basin could be illuminated in a variety of colors from below, so that the scene constantly changed.

By comparison, the Paris Exposition Universelle had just seven thousand lights of ten candlepower on the Eiffel Tower—only enough to limn its features, and there were only fifty-seven hundred incandescent lamps for the Palace of Electricity (see figure 5.5).[72] Exterior lights only outlined the buildings and emphasized their general architectural features—a practice common in 1890, but passé by the turn of the century. Stieringer's lighting in Buffalo created the equivalent of an electrified three-dimensional impressionist painting. C. Y. Turner, his "director

5.5 Eiffel Tower Illuminated, Paris Exposition, 1900
Source: Prints and Photographs Division, Library of Congress, Washington, DC

of color," assisted him. The two abandoned the "white city" ideal that had been the hallmark of the Chicago Exposition. Instead, as Robert Rydell explains, "Visitors to the Pan-American exposition stepped into a carefully crafted allegory of America's rise to the apex of civilization. The color mosaic presented by the fair told the story of the nation's successful struggle with nature and forecast a future where racial fitness would determine prosperity."[73] The color scheme expressed a vision of moral order. Turner explicated the color scheme: "As we enter the grounds from the park," one first encountered crude colors that expressed "elementary conditions, that is, the earliest state of man."[74] He used strong primary colors in this area of the fairgrounds, which housed anthropological exhibits that included Africans, Native Americans, Samoans, and Javanese, dressed in their traditional clothing, and living in what then seemed premodern simplicity or barbarism, but seems to have been exploited subjugation. One photographic book memorializing the fair characteristically declared that the Javanese were "a simple and happy people," although they "are gradually learning to be more in touch with people outside" their island.[75] As fairgoers moved further into the grounds, Turner's colors became "more refined and less contrasting" until one reached the tower at the center, which suggested "the triumph of man's achievement" and therefore was "the lightest and most delicate in color."[76] In this central area were the technological displays, with the Electric Tower at the apex. At its summit was a powerful projector, whose beams could be seen twenty miles away.

The electrical displays were reflected in the many canals and central basin. After first enjoying the view from the ground, many visitors took the elevator to a restaurant two hundred feet above the central court, where they could dine while viewing the spectacle. The Pan-American Exposition fused architecture, urban planning, and lighting into a coherent design that expressed an ideology of humanity triumphing over nature. Advanced technology was the measure and proof of that achievement, and the visitor moved to the center of that vision by walking into the grounds. Most fairgoers arrived at the main entrance by railroad or electric streetcar. This ensured that they proceeded from crude to delicate colors, from primitive life toward the spectacular vistas of technological civilization epitomized by the Electric Tower. Surviving black-and-white photographs cannot show the pastel shades that awed and charmed the crowd. Nor can they convey what one observer called the seeming "transparency, an airy, unsubstantial appearance" that gave the buildings a gossamer lightness and delicacy (see figure 5.6).[77]

Stieringer carefully orchestrated the moment when the displays were switched on. He allowed the entire fairgrounds to fall into obscurity as the sun set and the crowd gathered for the performance. Then the first electrical lights came on at the top of the four-hundred-foot-high Electrical Tower and spread out from that point like an artificial dawn. The tower's forty thousand small bulbs first gave it an ivory hue, with tints of blue, green, and gold. As the light swept down from the top of the tower and out to the other buildings on the grounds, the crowds gasped at the transformation of the site, which achieved an entirely new appearance. The buildings became more than flattened outlines or idealized patterns, being filled in like a pointillist painting, using delicate traceries of light to create a three-dimensional scene. It was an animated performance, not a static image. A young woman who witnessed the illuminations many times noted that "the light comes on by degrees, and this creates a novel effect." First, on the tower, "there is a faint glow of light, like the first flush which a church spire catches from the dawn. This deepens from pink to red, and then grows into a luminous yellow." Likewise, the other buildings could assume several different appearances, as electrical

5.6 Central Courtyard, Pan-American Exposition, Buffalo, 1901
Source: Danish National Library, Copenhagen

lighting was used for a succession of special effects. The daylight exposition "vanished and in its place is a wondrous vision of dazzling wonders and minarets, domes, and pinnacles set in the midst of scintillating gardens," animated by electric fountains.[78] Another commentator remarked, "When the current is a quarter from full, there always comes an intensely dramatic pause, like the rest for a deep breath that a great actor takes before striding to the footlights for his final and convincing flight. It is then that the applause is given, always the same and always spontaneous, to lapse again into silent admiration as the full glory is revealed. The color of the buildings, so radiant by day, is enhanced at night."[79] *Scientific American* had predicted in 1900 that the Pan-American Exposition would be "an unparalleled electrical triumph—a brilliant celebration of electrical development and achievement."[80] Visitors concurred. The *Cosmopolitan* complained that well after opening day, many exhibits were not yet ready, but it called the fairgrounds at night "a veritable fairyland" and a "triumph" achieved by "the masters of modern science over the nature-god, Electricity."[81] Edison expressed his enjoyment less bombastically, telling Stieringer and Rustin, "This is out of sight!"[82]

Every location was worth a second visit after dark, when it was transformed. The gardens and lagoons had "concealed electric bulbs" that made "every blossom to stand out as clearly as if under the rays of the sun. The inky blackness so common to lagoons has been overcome in the great expanse of the Court of Fountains by means of floating lights, which look like stars in an inverted sky, and the result is a luminous lake of golden fluid, agitated by the many fountain jets."[83] Amusement parks, hotels, and other businesses later copied these innovations. As techniques improved, designers combined lighting with an ensemble of other electrical technologies, such as projectors on the sky and electrical fountains that used an array of filters to change the color of the water, which alternately frothed, surged, and shot high into the air.[84] There had been a steady progression from gas lighting at early expositions to the powerful but blatant arc lighting of the Columbian Exposition to the more pointillist effects of thousands of colored lights at the Pan-American Exposition. Lighting had evolved from a small, crude demonstration to a vast display of color ornamentation and effects. By 1901, illumination had become a defining element in exposition design, capable of subtle colors, animations, coordination with musical accompaniment, and many special effects. US exposition lighting had advanced well beyond the European competition and put tower lighting systems in the shade.

Nevertheless, the social utility of the great expositions was declining. From 1851 until the end of the century, they had been useful educational institutions where manufacturers presented new products. But Stieringer felt that by 1901, the fairs "no longer meet a want in the way of enabling the public to become acquainted with what is latest and best in any art or industry. This work is now done very satisfactorily by the daily and the technical press." Furthermore, any large city "itself is likely to contain as good a display" as an exposition. Instead, Stieringer noted, "Expositions tend more and more to the development of the spectacular, and are becoming, unfortunately, overridden by the Midway feature."[85] Across the country, exposition spectacular effects were being assimilated into the urban commercial landscape.

6

Commercial Landscape

After 1890, two approaches to urban lighting battled for dominance. One espoused the aesthetics of expositions and encouraged lighting orchestrated to create a harmonious overall effect. Its advocates were often critical of skyscrapers, preferring a horizontal city where buildings had no more than five floors, as in London and Paris. They sought to restrict or eliminate large advertising signs. They did not want to retain gaslight but instead wanted to place electric street lighting on tasteful poles that harmonized with the surrounding architecture. Members of this group were well educated, and some were scions of wealthy families. They organized the City Beautiful movement, and influenced the design of the civic events and world's fairs examined in chapter 7.[1]

The other approach was individualistic and commercial. New US cities were laid out as a checkerboard of empty lots awaiting development, and owners wanted to maximize profits and property values.[2] Maximization frequently meant taller buildings, brighter lights, and eye-catching advertising. Taken to its logical conclusion, the result was a commercial district with a jumble of signs and no architectural harmony. Each business sought as much attention as possible, culminating in the frenetic displays of Times Square. This commercialization of night space prevailed in the United States more than in Europe. London permitted electric advertising in Piccadilly Circus and a few other sites, but restrained it elsewhere, and in the 1890s, even in Piccadilly, it was modest compared to New York's Broadway. Paris had a similar policy, and bathed its historic buildings along the Seine in white light. Venice provided an even stronger contrast. As a pedestrian city without automobiles or skyscrapers, it never adopted intense public lighting, but installed a few streetlights, supplemented by light falling from the windows of residences, shops, and public buildings. Venice remained a walking city, with intimate, small-scale illumination and few electric signs. Yet this should not suggest a simple dichotomy between Europe and the United States. The lighting in Berlin, for example, resembled that in New York more than London.

Until the late eighteenth century, British businesses hung simple signs over their doors and used generic symbols to indicate their trade. In the late eighteenth century, though, "whimsical shopkeepers" joined "heterogeneous objects joined together," such as blue boars, flying pigs, the "Lamb and Dolphin," or "Three Nuns and a Hare."[3] A French person visiting Britain in 1765 was surprised at "the enormous size of the public house signs," and observers remarked on "many gross errors" in spelling and grammar.[4] By 1800, British shops commonly had signs that covered "the whole front of a house" with letters three feet high. A clothing merchant in Cheapside had "a prodigious grasshopper" sign that attracted so many customers that other merchants copied it and paid a fee for hanging similar grasshoppers on their shops. Displays also extended to "coaches and sedans of the wealthy classes" that were veritable "picture galleries, the panels being painted with all sorts of subjects." Some theaters paid as much as £500 to have professional painters produce signs, including one depicting William Shakespeare that hung in Drury Lane. A few coach painters were even members of the Royal Academy. This extravagant phase of sign production ended with an "Act of Parliament for removing the signs and other obstructions in the streets of London."[5]

Illuminated gas signs emerged after 1810. In 1817, one of the first gas installations at a Parisian café spelled out in flaring letters "café of hydrogen gas."[6] Gas signs were boxes "with the gas jets inside, … with varicolored glass jewels outlining the letters." These signs were common outside drugstores, oyster houses, and shops. Locations inside or sheltered from the wind might have "brass or copper tubing bent in the shape of letters and drilled at regular intervals with the gas pressure so regulated so that only a tiny flame was emitted."[7] These signs could spell words or represent objects in a range of colors, and in the United States they were still being manufactured as late as 1900.[8]

The entertainment industry realized lighting's potential for space making. When Paris theaters installed gaslight, they changed their previous practice of lighting the hall throughout a performance. After 1822, they heightened the distinction between the stage and auditorium by turning down the gaslight when the performance began, thereby focusing attention on the stage and heightening the power of illusion. Parisian shops took note, and their windows gradually expanded in size and adopted dramatic lighting, as though they too were stage sets.[9] By the 1830s, shops were often lighted by gas burners outside and above their windows, using reflectors to illuminate the goods. Beginning in the 1850s, large plate glass windows replaced many small panes, and displays became larger and more elaborate. While these changes occurred in all large cities, Europeans built many arcades, in Paris and London, of course, but also in Leeds, Mainz, and other regional centers. Fewer were built in the United States.

Early shop windows contained mostly handmade goods. Factories were few, and brand names scarcely existed. Newspaper advertisements consisted mostly of text, with small, simple drawings. Gaslight illuminated shop windows or drew attention to the entrance of a building, but it seldom burned after closing. Few billboards were illuminated by gas, and posters and painted signs were the norm. By the late 1860s, 275 US bill-posting businesses plastered fences and walls with temporary advertisements.[10] The public unveiling of the first "sixteen sheet billboard in Cincinnati in 1878 attracted crowds" so large that police were called to keep order.[11] An expert on billboard advertising declared in 1896, "[It] compels attention; it stares people in the face at every turn. They cannot escape becoming more or less familiar with the things advertised if they look about them at all." Anticipating what became common practice with electrical signs, he argued, that "a downtown retailer who caters to the whole population of the city could use both the billboards and newspapers to advantage—the former to call people's attention to his store … and the latter for a description of goods and prices."[12] In 1905, on an average day in Detroit there were 65,000 bills on the hoardings.[13] "An estimated fifty miles of billboards edged the city's streets" in Chicago, and in 1908, "more than 8.5 million linear feet of billboards lined the nation's sidewalks, streets, railways, and roadways."[14]

Illuminated signs were rarities compared to posters, but drugstores, tobacco shops, saloons, and theaters used them. In the 1840s, Barnum's Museum on Broadway in New York had one of the country's largest gaslight signs, and the front of his building was covered with "five-foot high transparencies depicting over a hundred species of animals."[15] Cities adopted gas signs for celebrations and special occasions. For example, "displays fitted with gas jets welcomed home Philadelphia troops from the Mexican War in 1848," including a thirty-foot Goddess of Peace.[16] By the 1850s, many businesses used gaslight to draw attention to their location.

Gas technology continued to be used in signs for the rest of the nineteenth century, but it was less versatile than electric light. It could create an image, yet did not lend itself to the illusion of movement. Using shades and reflectors one could get some effects and colors, but electric bulbs had a wider spectrum. Gradually, almost every business, however small, adopted electricity not only in its show windows but also to brighten its facade, emphasize its name, and operate a sign. A jeweler might use tightly focused lights to highlight gemstones against a background of black velvet. A hotel might present itself as a quiet and safe oasis or center of entertainment with a ballroom, bars, and restaurants. By establishing expectations, lighting could draw or repel potential customers. Each business developed its visual vocabulary, combining a logo, characteristic colors, brand name, and architectural features into an overall definition of its site. These individual efforts had a collective impact quite unlike the moonlight towers or stately visions of the world's fairs.

In the 1880s, it became evident that US stores lighted with gas attracted fewer customers than those equipped with electrical lights, which spread rapidly. Commercial lighting emerged piecemeal with no coordinated intent and gradually coalesced into unintended visual effects. The vista of a world's fair was carefully planned, but the signs on a US city street haphazardly juxtaposed competing styles and messages. By the early 1880s, these individual efforts began to have a larger effect. As one guidebook explained in describing New York's Sixth Avenue, by day its sidewalks were thronged with shoppers, "the best of New York's people, intent upon honest business." Yet "when darkness settles down over the city, and the lamps flare out along the street, and the broad rays of light

6.1 Illuminated Sign Advertisement, 1901
Source: Hammer Papers, Warshaw Collection, Smithsonian Institution Archives, Washington, DC

stream brightly into the open air from the stores, restaurants, and saloons, Sixth Avenue undergoes a transformation." The large stores closed, but many smaller shops, "the saloons, restaurants, and tobacconists," stayed open and gave "a brilliant coloring to the street with their bright lights and elaborately decorated windows." Overhead were "the many colored lights of the Elevated Railroad" that further animated the scene with "the roar of the brilliantly-illuminated trains."[17] There were two Sixth Avenues: by day sober and honest, but by night a raffish mixture of ordinary citizens, streetwalkers, pickpockets, and ruffians.

As stores remained open longer and nightlife intensified, the perception of darkness changed. The explicit message of world's fairs and trumpeted goal of reformers was that darkness be vanquished as part of the imperial movement of civilization. Just as the Enlightenment reconceived madness in order to define the "age of reason," the Victorians redefined darkness as part of their project of intensifying illumination. The central city sparkled in contrast to shadowy slums, which became perversely attractive zones of illicit pleasure. In New York, the Great White Way was all the more brilliant juxtaposed to the Bowery. Already during the gaslight era, guidebooks stressed such differences. In 1869, Matthew Hale Smith's *Sunshine and Shadow in New York* described the great hotels, homes of the rich, and vistas of the city, and declared, "Broadway is a perpetual panorama. Its variety never tires." But it also devoted twenty pages to prostitution, "the social evil in New York," and even more space to gambling houses. This was a city of extremes. Nowhere was philanthropy greater, and yet "the base men of every nation, and the crimes, customs, and idolatries of every quarter of the world are here." It seemed that "great cities must ever be centers of light and darkness."[18] As lighting intensified, so did this duality.

Intensive lighting sharpened awareness of class difference. Expanding cities developed distinctive neighborhoods, with the best lighting in the most prosperous areas. The contrast made districts such as New York's Tenderloin or Chicago's Levee seem exotic, and the curious went "slumming."[19] Some felt that the authentic city emerged in darkness, and thought encounters with vice and the criminal class revealed the "real" city. Small groups paid guides for nocturnal tours, where they might glimpse gamblers, pickpockets, ladies of the night, drunks, and con artists.[20] For those who dared not visit lower-class dance halls, gin joints, brothels, illegal dogfights, homosexual "masquerades," or black ghettos, books catered to their prurient interests. In *Maggie: A Girl of the Streets*, Stephen Crane took a sympathetic view of his heroine in the slums, but some guidebooks and popular novels depicted the poor as savages living in an urban jungle.[21] Reformers wrote exposés, perhaps most famously Jacob Riis's *How the Other Half Lives.*[22] Riis went into the poorest sections of New York, first as a journalist, then as a photographer and reformer, capturing images of squalor, suffering, and poverty that he displayed to genteel audiences in public lectures. Fears of the impoverished encouraged ever more powerful lighting as a deterrent, giving the police greater powers of surveillance. Yet every increase in lighting made the dark parts of the city more mysterious.

Even as this dichotomy intensified, large-scale spectacular advertising began to emerge in city centers. Before monumental electric signs could emerge, however, three things were necessary: large corporations, brand names, and mass production of such things as cigarettes, biscuits, and chewing gum. Most production was small scale until 1860, and markets were regional. Few goods were nationally known. It was not possible to register a brand name until 1871. Before then, if one wanted salt or oatmeal, the grocer sold it by weight from a large sack. By the 1890s, however, Quaker Oats and Morton Salt had established national brands through advertising and distinctive packaging. Likewise, until industrial-scale canning developed during the Civil War, milk and pickles were local products, but soon they too had national brands, notably Heinz and Carnation. Companies that mass-produced a brand name product began to erect enormous electric signs on Broadway, like one showing a giant green Heinz pickle. Another depicted a girl carrying an umbrella who perpetually spilled Morton Salt, under the slogan, "When It Rains It Pours." Before brands existed, advertising budgets were small. In 1869, New York had 42 advertising agencies, most of them just a few people in an office; two decades later there were more than 280 agencies, and they had grown in size and scope.[23] Where once advertisers did little more than buy space in printed media for clients, they had begun to write copy, create images, establish slogans, and orchestrate national campaigns.

One might assume Americans pioneered electric advertising and exported it to Europe. But an English visitor to New York in the late 1890s found that "sky signs" were "unknown in New York, so are the flashing out-and-in electric advertisements which make night hideous in London," particularly "the illuminated advertisements of whiskey and California wines that vulgarize the august spectacle of the Thames by night." In New York, "two large steady-burning advertisements irradiate Madison Square," which this visitor found inoffensive because they did not flash.[24] In short, the English independently had developed garish electrical displays. Yet just as they earlier had suppressed enormous painted shop signs, they restricted electric advertising.

Before 1900, the sale of electric signs was driven by consumer demand with little stimulation from utilities. At meetings of the National Electric Light Association, managers seldom discussed electric advertising before 1902. In that year, 10 percent of all incandescent light business in New York City was due to electric advertising, even though the signs primarily used small bulbs of eight candlepower. Many signs used flashers based on the one Hammer had designed in 1883 and displayed at the Health Exposition in Berlin.[25] Later, flashing signs spelled out words, letter by letter, or created the illusion that "an invisible pen" was "tracing the letters in lines of fire." Flashers could depict a rippling sea, constantly waving flag, or illusion of circular motion.[26] There were also "signs operated by keyboards that permit the instantaneous spelling of any desired word or sentence; detached metal letters of all shapes and designs, which permit the consumer readily to make up any legend; decorative signs of all shapes; sockets joined with adjustable links, permitting the rapid formation of any letter or design; electric-lighting boards on which any combination of letters or design may be traced and then outlined electrically by implanting bulbs."[27] One executive declared, "Taken altogether [the signs] give to the thoroughfare a wonderfully high degree of attractiveness." They had "a double value. The first is advertising, the second light; both are desirable." Electric signs ultimately advertised not products but themselves. Installing one sign on a dark block almost invariably led to more signs and brighter show windows. A new night landscape emerged: "Where is such another sight to be found as the vistas on our Broadway, looking in almost any direction in the evening hours," except in other large US cities?[28]

The National Electric Light Association surveyed more than a thousand utilities in 1905 and estimated there were seventy-five thousand electric signs in the United States, thirty thousand of them installed during the previous year.[29] The market was unevenly developed. A few towns of less than ten thousand people had more than a hundred electric signs, yet some cities of fifty thousand people had only two or three. Even so, electricity sales for signs were $4.2 million a year—a figure that better marketing could quadruple. Small businesses owned the majority of these signs, buying electricity on a multiyear contract, paying a flat rate. Such signs were usually turned on automatically for four hours a night. Many utilities supplied at no cost a basic sign that spelled the name of a business or product.

The largest, most distinctive signs were built on steel frames and handcrafted by specialized firms. A certain percentage of the surface area had to be empty to allow the wind to pass through, or otherwise the pressure might topple it. Because they were legible only at night, most of these signs were in central locations where a large public passed by, such as the Loop District in Chicago, Atlantic City Boardwalk, Market Street in San Francisco, or Campus Martius in Detroit. These signs became local landmarks. Along Broadway, there were always twenty-five to thirty enormous signs erected by national corporations, advertising mass-produced goods such as salt, soap, cigarettes, breakfast cereal, razors, and chewing gum.[30] As an early textbook on outdoor advertising emphasized, "No form of advertising carries more prestige or makes a quicker or more lasting impression than electric display. … [I]t exerts an influence which is extremely powerful."[31]

There were also distinctive signs for businesses unique to each city, notably department stores like Macy's in New York or Filene's in Boston. In Chicago, a restaurant on North Clark Street suggested that one "Eat, Night and Day." A bar with scantily clad dancers challenged passersby with a sign that said, "Don't Look." A large sign displayed a fountain where white water shot up from the base and fell back in a graceful curve; red lights were then added in a pleasing color display. Chicago's children liked a sign where an electric mouse appeared and ran around the sign until jumped and chased by another mouse, and then another, until they saw "a whole string of mice playing 'follow the leader.'" It "looked like a marathon race for bleached rodents," until the mice scurried back into the darkness. A Chicago amusement park's elaborate sign depicted a sky-rocket shooting up and exploding into hundreds of colored lights that slowly floated down; a billiard parlor's sign showed a giant white ball that ran into a red ball and then a second white ball, scoring a point. "Then the balls roll on into position for the next shot." Another sign depicted "a tremendous fireplace" where lights simulated dancing flames.[32]

Growing crowds of white-collar workers were out at night, and between 1900 and 1910 it was not unusual for the then-silent films to be projected outdoors in city centers (see figure 6.2). Richard Harding Davis noted that in the evening, the New York "clerk appears, dressed in his other suit, the one which he keeps for the evening, and the girl bachelor … puts on her hat," deserting her cold-water flat to join "the unending processions on Broadway."[33] Their working hours had decreased and their disposable income had risen.[34] Throngs patronized vaudeville, music halls, theaters, dance halls, and cabarets, all of which used scintillating lights to lure in customers. As David Nasaw observes in *Going Out*, white-collar workers were a crucial part of the audience for electric

6.2 Times Square Crowd, Outdoor Movie Projection, 1908
Source: Prints and Photographs Division, Library of Congress, Washington, DC

advertising. Between 1880 and 1910 the nation's clerical workforce increased 1,000 percent, from 160,000 to 1.7 million, and was concentrated in cities.[35] Joining these clerical workers were at least as many blue-collar ones whose hours were becoming shorter, while their real wages rose. Between the Civil War and 1900 the percentage of income spent on food dropped from 67 to 43 percent, freeing up discretionary income for entertainment.[36] In the last decades of the nineteenth century, more people went out in the evenings, and more of them were unchaperoned women. (Many people also worked into the night under artificial lights. They labored in steel mills, cleaned office buildings, worked in restaurants, served as hotel porters, drove streetcars, repaired subways, printed newspapers, operated railroads, and carried out a myriad of other tasks.) When off work, millions ventured into the new metropolitan nightlife. They usually went as groups, not as solitary flannêurs, visiting dance halls, vaudeville shows, skating rinks, and amusement parks. It was not necessary to buy a ticket in advance, as most entertainments were continuous. A new song began at the dance hall every few minutes, kinescope parlors had many short films for individual viewing, and vaudeville houses and the first movie theaters ran their shows continuously. Going out on the town, once a luxury, became commonplace. Young men out on a spree had long wandered the night city, but now women also went out, particularly to dance halls, shopping districts, and amusement parks.

Corporations realized this evening crowd was an advertising opportunity. When Dreiser arrived in New York in the mid-1890s, he saw the Madison Square Garden Building, designed by McKim, Mead, and White, with a 300-foot tower lighting up the night. At the top stood a 14-foot sculpture of the huntress Diana on a crescent moon, with ten powerful lights at her feet. Augustus Saint-Gaudens created her, and though she weighed 1,800 pounds, Diana rotated easily on ball bearings and became the city's largest weather vane.[37] The building was inaugurated in 1891 with pyrotechnics, and for 25¢, it offered a spectacular view of New York from the top. "The entire structure was illumined by the glow of its 8,000 incandescent lights, its 100 arc lights, and its search lights. The effect was such as to bring all passers-by to a standstill."[38]

Dreiser also was struck by an enormous advertisement on the site where the Flatiron Building was later constructed. A sign 50 feet high, 80 feet wide, and covered by light bulbs announced that Manhattan Beach was "swept by ocean breezes," had "three great hotels," and offered as entertainment "Pain's Fireworks," a racetrack, and "Sousa's Band." The sign was not automated. A worker on a nearby roof turned a switch to illuminate its seven lines, each in a different color. He then turned all of the lights on at once before repeating the process. Such signs were soon automated using a commutator that worked much like a mechanical music box. A revolving metal drum punctured with holes turned on and off different combinations of lights as it opened and closed electrical connections. Heinz rented the space in 1892 and erected a larger sign, featuring a 45-foot pickle in green bulbs as well as vivid red reminders that the corporation also sold ketchup and other condiments (see figure 6.3).[39] In little more than a decade, such large signs appeared in every large US city, animating the night. A few companies specialized in making them, notably O. J. Gude in New York. Between 1890 and 1918, his workshop produced many famous signs including the one Dreiser saw. Regarded as a philistine by some, Gude raised the artistic level of advertising and eventually joined the Municipal Art Society, where he proposed that some billboards be turned into outdoor art galleries. He asked his employees to study art and believed that advertising could be a source of uplift.[40]

Gude knew that electrification permitted lighting arrays and special effects more powerful than those possible with gas. Gaslight was weaker than incandescent bulbs. Gas installations could not be blinked off and on to draw attention. Gaslight could give off different colors if placed behind colored glass, but the range of hues was limited compared to those available in electric bulbs. The wind disturbed gaslight, and even when shielded by glass, it moved and flickered. Electric lights were steady, and came in many sizes and wattages, making possible precise designs. As sign makers learned to exploit these advantages, animated arrays could depict a chariot race, a person driving a golf ball, or a diver plunging into a pool of water. In Times Square, a popular advertisement for underwear depicted a brief boxing match. Another portrayed two gigantic toothbrushes, "and hanging on the bristles of them a little devil, little but gigantic, who kicks and wriggles and glares. After a few moments the devil, baffled by the firmness of the brushes, stops, hangs still, rolls

6.3 Heinz Electric Sign, ca. 1900
Source: New York Public Library

his eyes moon-large, and in a fury of disappointment goes out."[41] One illuminating engineer noted in 1910 that the "most elaborate signs … are almost the equivalent of an entire vaudeville act. In fact, many of the schemes used in the theater to produce motion effects are utilized in electric signs," such as making wheels go around on a vehicle while "the nearby objects pass by." He described an automobile advertisement depicting "a joy ride. The dust is seen flying from the revolving wheels, the smoke from the men's cigars floats away and the ladies' veils flutter in the breeze." The sign even contained an electric horn that honked.[42] Millions of visitors to Times Square viewed these "electric vaudeville acts," experiencing them individually and as an overwhelming totality where individual messages were lost in the visual cacophony, for it was impossible to process all the signals bombarding the senses.

A few of the largest signs did not sell a product but, rather like a television program, offered entertainment that several companies sponsored. The most famous example was an enormous sign made by the Rice Electrical Display Company of Dayton, Ohio, and erected on the roof of New York's Hotel Normandie in 1910. Seven stories high, the sign portrayed a Roman chariot race, using 20,000 bulbs, 70,000 connections, and 2,750 switches. The apparently revolving chariot wheels were 8 feet high. The sign showed not only the rapid movement of the chariots themselves but also the drivers whipping their horses, flying of the horses' manes, dust thrown up behind, and crimson robe of the driver flying backward. It told a story, and its performance lasted several minutes. Afterward it flashed the names of sponsors, including Armor and Co., Remington Typewriter, Prudential Insurance, National Cash Register, and Quaker Oats. Each night enormous crowds saw the chariot race, and most stayed for a second performance.[43] The sign only came down because a new building obstructed the view.

Electric signs were concentrated in business districts, on major streets, and especially in squares. In Boston, a clothing store covered its exterior with lights, including a 6-foot-high shamrock and electric American flag.[44] Chicago alone had 2,000 electric signs in 1905.[45] In a few locations, such as Times Square, the advertising lights were so numerous that there was little need for street lighting, such as beneath the enormous Wrigley's sign in Times Square (see figure 6.4). But this

6.4 Wrigley's Sign, Broadway, c. 1920
Source: New York Public Library

was exceptional. As the National Electric Light Association explained in a booklet, "Properly placed, electric signs draw people to a street, particularly if they tower high above some roof and are seen from a distance. But the roof signs will neither illuminate the street nor induce people to pass directly by the particular stores over which they are mounted." Alternately, "if they are placed low enough to illuminate the street they cannot be seen from a distance. Hence they lose in advertising value." In short, "electric sign lighting cannot take the place of street-lighting." This realization led local businesspeople to band together and pay for improved street lighting, often called white ways, in downtown Wichita, Cincinnati, and other cities.[46] Los Angeles merchants did likewise, and in May 1905, celebrated their new ornate lampposts in a parade that included floats, "bands and tallyhos."[47]

After streetlights were installed, every store seemed comparatively dark, and needed to upgrade its show window lighting and brighten its signage. Otherwise, its goods were lost in shadows. All three kinds of lighting were essential. Pole lights along the curbside attracted "people to a street; electric signs emphasize certain stores or buildings; window lighting leads to the inspection of goods. … Each method helps the other."[48] The expanding market soon included noncommercial buildings too, because they faded into the background if they adjoined brightly lit businesses. Even churches invested in exterior lighting.[49] Some erected lighted crosses, including churches in New York's Bowery and Washington Square. The Anglican Cathedral in Denver rented space on a billboard.[50] The National Electric Light Association urged utilities to copy successful campaigns in Worcester, Cleveland, and San Jose, which convinced church vestries that attendance soared when stained glass windows were illuminated.[51]

The largest advertisements exploited the place-making potential of electric lighting to create environments that lacked spatiality: they could be seen but not entered. The individual could never be "from" or "go" there. Such a "place" could only be viewed. Moreover, the signs in Times Square and other centers of advertising were all on different scales. A human figure might be much larger or smaller than life size. A bottle of ketchup could be taller than a house. Such commercial places repeated sequences of brilliant effects, imposing a rhythm and pattern of light

6.5 Atlantic City Boardwalk at Night, 1910
Source: Prints and Photographs Division, Library of Congress, Washington, DC

and shadows on their surroundings. Progressive era reformers had called for new urban spaces such as museums, parks, playgrounds, and squares "as a context and shaping force of public life."[52] They would uplift and educate the public, which occasionally could be inspired by an evening parade. But the landscape of advertising was permanent. It proclaimed a unity based not on history, education, or active participation but rather consumption. The places that advertising created were fragmented, and the displays impermanent; they shaped not active citizens but avid consumers. This process culminated in elaborate Christmas lighting that was already customary by the 1890s.[53] The spectacles of outdoor advertising made absolute the city's heterotopian multiplicity.

Nowhere was the landscape more fully electrified than in the new amusement parks near every US city, epitomized by Coney Island and Atlantic City (see figure 6.5). Even if located in places without much intrinsic scenic attraction, at night they proved exceedingly attractive. Amusement parks were particularly popular between 1890 and 1920, attracting crowds of young people intent on a good time. They appealed to the middle class as well as the rapidly growing class of clerks, secretaries, shopgirls, and other workers. The parks originated in expositions, notably in Chicago in 1894, where the Midway with its exotic dancers, beer halls, curiosities, games, and rides drew visitors away from the official exhibits. Amusement parks often emerged at the seaside resorts such as Coney Island and Atlantic City in the United States, or Blackpool and Brighton in the United Kingdom. Already in 1879 Blackpool had electrical generators and a system of arc lighting that extended operations into the evening.[54] Its US cousins did the same. Amusement parks mushroomed in the 1890s, after electric streetcar lines were built across the United States. These lines owned power plants that were nearly idle at night and on weekends, until they built amusement parks whose electrified rides and extravagant lighting could run on the excess generating capacity. As one consultant declared, experience showed "there is nothing like brightness in a summer park, and you cannot overdo it."[55] The twenty-largest US cities all had several amusement parks. They were also located on the outskirts of smaller cities, such as New London, Schenectady, Harrisburg, Columbus, Muncie, Dubuque, and Portland (see figure 6.6).

6.6 Luna Park at Night, Coney Island
Source: Prints and Photographs Division, Library of Congress, Washington, DC

The parks offered escape from convention. Visitors dressed more casually than in the central city, and combined the uninhibited pleasures of sun and sea with roller coasters, fun houses, and intimate boat rides through the tunnel of love. Parks developed a dynamic, vernacular aesthetic that mixed the extravagant use of small light bulbs pioneered at the Omaha and Buffalo expositions with the riotous color and vigorous advertising of the Great White Way. There were literally millions of lights at the largest sites such as New York's Coney Island or Philadelphia's Willow Grove Park. The Philadelphia Rapid Transit Company ran no less than seven streetcar lines to the park's entrance. Willow Grove had a roller coaster, "mountain scenic railway," two carousels, and other electrified rides. At the center of a lake, with a restaurant and café nearby, stood an enormous electric fountain that cost \$100,000.[56] The gaudy commercial aesthetic of such parks offended social reformers. Yet as John

Kasson has argued, these egalitarian sites broke down barriers between ethnic groups, and appealed to both the poor and middle classes. When visitors were thrown together on rides and crowded beaches, social distinctions temporarily fell away, with the significant exception that blacks were usually excluded.[57] For those admitted, the elaborate electrical displays helped create the sense of entering a liminal zone that suspended social hierarchies and norms. Park designers discovered that individuality disappeared in the crowd, and it was profitable to pack attractions together. One expert advised, "The greater the area, the less chance of making money. Concentration is the thing. Keep the people together in a bunch, the larger the crowds and the greater the inconvenience, the better they like it, and the more easy it is for them to spend their money."[58] These lessons were not lost on other businesspeople, and the mass culture emerging in city centers adopted lighting much like that at amusement parks. Theater marquees, flashing lights, and advertising signs increasingly dominated the night cityscape.

With such varied and intensive lighting, the city transcended previous experience, and the differences surprised Europeans. In 1904, the Englishman Philip Burne-Jones declared, "Broadway at night, with its myriad brilliant lamps, the names of its theatres and restaurants picked out in blazing points of electric fire, is a sight not readily to be forgotten, and one which impresses itself upon the imagination as much as anything in the great city." He preferred skyscrapers by night. "It is only at night that these buildings are tolerable. Then, with the electric lights gleaming from a hundred windows, the dark mass of their giant forms silhouetted against the evening sky, there is something weird and fantastic about them which appeals to one—something strange and characteristic, if not actually picturesque. At all events, they are unlike anything else on earth."[59]

Another British visitor exclaimed that

> under the purple, star-lit sky, street life in the central region of New York is indescribably exhilarating. From Union Square to Herald Square, and even further up, Broadway and many of the cross streets flash out at dusk into the most brilliant illumination. Theatres, restaurants, stores, are outlined in incandescent lamps; the huge electric trolleys come sailing

> along in an endless stream, profusely jeweled with electricity; and down the thickly-gemmed vista of every cross street one can see the elevated trains, like luminous winged serpents, skimming through the air. The great restaurants are crowded with gaily-dressed merry-makers; and altogether there is a sense of festivity in the air, without any flagrantly meretricious element in it, which I plead guilty to finding very enjoyable. From the moral, and even from the loftily aesthetic point of view, this gaudy, glittering Vanity Fair is no doubt open to criticism. What reconciles me to it esthetically is the gemlike transparency of its colouring. Garish it is, no doubt, but not in the least stifling, smoky, or lurid. The application of electricity—light divorced from smoke and heat—to the beautifying of city life is as yet in its infancy. Even the Americans have scarcely got beyond the point of making lavish use of the raw material. But the raw material is beautiful in itself, and in this pellucid air (the point to which one always returns) it produces magical effects.[60]

This description from 1899 predated most of the animated signs that soon dominated Broadway and made it even more lavish. By 1912, British advertisers were in talks with a leading US electric sign maker, who remarked that compared to Broadway, "Trafalgar Square has no night life" and "Piccadilly Circus is too small."[61] In contrast, electric advertising expanded rapidly in Berlin, particularly on Friedrichstrasse, which a *New York Times* reporter in 1914 thought "almost as brilliant with electric signs as Broadway in the region of Times Square, while Potsdamer-Platz, Berlin's Piccadilly Circus, is probably the most brilliant spot in the world at night."[62]

Berlin had no skyscrapers, however. While electric advertising spread rapidly to smaller businesses, theaters, dance halls, and sites of popular culture, corporations were slow to illuminate the facades of their office buildings. Early US skyscrapers added exterior lighting almost as an afterthought. Architects long resisted designing buildings with lighting in mind, and for decades illuminating engineers found it difficult to work with them. Even leading professionals such as General Electric's D'Arcy

Ryan continually met opposition.[63] In 1930, one prominent illumination engineer recalled, "Up to a few years ago, after nightfall, the building ceased to exist," disappearing into the darkness, with all its ornamentation invisible unless it happened to be illuminated by an advertising sign.[64] For a quarter-century, the illumination of skyscrapers from office windows was thought sufficient to create a nocturnal landmark. Buildings might add an illuminated clock, but the idea of turning the entire structure into a dominant object on the night skyline came surprisingly late. Only in 1907 did the 612-foot Singer Building, then the tallest in the world, make spectacular lighting part of its architectural plan. It also included an observation deck where, for 50¢, tourists could view New York City and its harbor.[65] Using thirty projector lights, it became the first skyscraper with a fully illuminated exterior. As at Niagara Falls, the design was by D'Arcy Ryan, General Electric's head of illuminating engineering, who for three decades after the death of Stieringer in 1902 was the leading lighting engineer. A 1908 photograph shows the Singer Building as a pillar of light soaring above Manhattan (see figure 6.7). The building dominated the skyline and became an advertisement for the Singer Sewing Machine Company. It elicited much social comment and was even the subject of a short story by Willa Cather.[66]

Owners of other skyscrapers immediately saw the publicity value of exterior lighting and illuminated their buildings as well. In contrast to the tower lighting of the 1880s, these were not sources of public lighting but rather displays of corporate power. The most notable example was the Woolworth Building, "New York's paradigmatic skyscraper of the early twentieth century."[67] Such Beaux-Arts skyscrapers wedded a gothic-inspired exterior that included gables, gargoyles, and finials to the latest methods of steel construction. In 1912, the Woolworth Building was the world's tallest, and lighting was carefully designed into the structure, including highly reflective terra cotta surfaces. The "objective was not only to bathe the tower in a sheet of light, but to emphasize the architectural details. Conventional searchlights would have transformed the surfaces into plain areas."[68] Instead, corrugated reflectors diffused the light. "The main architectural features of the mansard roof extending from the fifty-third to the fifty-seventh floor, the observation balcony at the fifty-eighth and the lantern structure at the fifty-nine and sixtieth

6.7 Singer Tower Illuminated
Source: Hall of History, Schenectady, NY

floor are covered with gold leaf. By the proper placing of projectors a glittering effect [was] obtained from these gold surfaces."[69] Under the direction of H. H. Magdsick, a General Electric lighting engineer, forty electricians worked on the project, which required 500,000 feet of cable and 550 powerful tungsten lights. The lights were turned on in 1913, impressing the spectators crowded into City Hall Park, along Broadway, and as far away as the New Jersey shoreline facing Manhattan.[70] The building was crowned with "lanterns in the crest of the spire. Twenty-four 1000-watt tungsten lamps were placed behind crystal diffusing glass" and appeared to be an "immense ball of fire" with a flickering effect caused by a dimmer that altered the light's intensity from a deep red glow to "flares of fifty times this intensity."[71] This light was visible fifty miles away (see figure 6.8).[72] From the tower, news of the opening was sent by wireless to the Eiffel Tower in Paris, and the Associated Press carried the story around the world.

The illuminating engineer Matthew Luckiesh called the Woolworth Building "a majestic spire of light … projecting defiantly into the depths of darkness … a torch of modern civilization."[73] The structure was soon nicknamed "The Cathedral of Commerce."[74] Its owner, Frank Woolworth, also installed playful lighting for special occasions. On July 4, 1914, strings of lights were draped "over the Woolworth tower's setbacks, creating the glimmering pinpoints of light characteristic of dazzling mass amusement schemes such as Coney Island's Luna Park."[75] He understood the advertising value of the building's illumination and saw it as a "giant signboard."[76] His building became the subject for many painters, notably John Marin, and magnet for tourists, who took a high-speed elevator to the observatory near the top. Subsequent skyscrapers were designed with their night appearance in mind, whether corporate headquarters like the Chrysler Building or landlords to a range of tenants like the Empire State Building. Skyscrapers became premier sites of corporate image making. Thirty years earlier, US cities had installed tower lighting to illuminate entire communities. In 1910, they celebrated private towers that illuminated only themselves.

US streets developed no common aesthetic. Each skyscraper and electric sign had its own form, and any harmonies that resulted were happenstance. Yet this eclecticism could be attractive when viewed from

6.8 Woolworth Building, 1913
Source: Prints and Photographs Division, Library of Congress, Washington, DC

a distance, whether one observed the skyline or the flashing lights of the commercial zone. The whole seemed more than the sum of its parts, even if classical standards of symmetry, proportion, and a common architectural vocabulary had been abandoned. The exuberance and unpredictability of spectacular lighting relieved the monotony of the grid street pattern. Urban tourists flocked to skyscraper observation decks for a panoramic vision that literally widened their horizons and turned the city into an abstraction. The world below, suddenly miniaturized, became an enormous pattern, a hieroglyphic for interpretation. To read this pattern required myriad realignments and reinterpretations of the scene. The visitor, having triangulated some of the spatial relationships known from walking through the scene below, could make sense of the vista without reconstructing every detail. The vista presented the enormous material complexity of modern society. Especially in the evening when millions of lights transformed the city, the view became an ineffable affirmation of the technological sublime.[77]

This scene was not entirely the product of commerce. The City Beautiful movement also had shaped some icons of this urban landscape, notably New York's Flatiron Building. The corner where the Flatiron stood was a battleground between the two visions of the electrified city. In the 1890s, the site was occupied by an unremarkable low building, behind which stood the high, flat wall that Dreiser had seen covered with flashing advertisements for Coney Island, and where Heinz later put his giant green pickle that was visible far up Broadway. Adherents of the City Beautiful movement disliked the Heinz sign and were pleased that Daniel Burnham, architect of the Chicago Exposition, designed the Flatiron Building, which blotted out the wall. The unusually thin edge of his steel-framed building became iconic. Edward Steichen photographed the Flatiron Building on a wet evening in 1904, shortly after it had been completed. His image suggests how the modern city was cut off from the stars and had reduced the moon to a secondary source of light. Steichen's impressionistic building is softly framed by a starless sky that is brightened from black to gray by the collective lights of New York that have diffused into the damp air. In his time exposure, the atmosphere is brighter than the building, whose lower stories are lighted only by the electric streetlights along Broadway. In the foreground, streetlights reflect off the wet pavement, where a man stands in the park, back to the camera, taking in the scene. The filigree of delicate branches cutting across the building contrasts with the straight lines and towering mass of the Flatiron.

Steichen captured one of the new visual pleasures that the skyscraper city offered. In the 1870s, New York had been far dimmer, and a walker would not have seen such a view. Almost no skyscrapers existed in the 1880s. In 1900, the electric Heinz pickle winked on and off against a vast orange background. It was visible a mile uptown and unceremoniously disrupted the stately white lighting that celebrated the victorious return of Admiral Dewey. The Flatiron Building replaced that sign, but it was a temporary victory. Larger and more spectacular advertisements soon lined Broadway, and new skyscrapers like the Singer Building splashed themselves with light. Steichen's photograph recorded a brief period when tall buildings remained dark at night, even as New York's light pollution brightened the sky. Like the Flatiron, the City Beautiful movement as a whole was in tension with the intensely commercial elements of US cities.

7

City Beautiful

To the City Beautiful movement, the commercialization of light expressed an individualistic social fragmentation that adherents worked to overcome by holding expositions, building public libraries, opening parks, establishing museums, and creating other uplifting institutions that would put all social classes on an equal footing. In such public spaces, lighting could make knowledge, culture, and entertainment accessible to all. City Beautiful proponents were disturbed by the commercialization of urban public space. As Peter Baldwin notes, the new white ways were "buzzing with energy and barely suppressed eroticism." Reformers such as Jane Addams feared that such districts might overwhelm the still-undeveloped sensibilities of adolescents. The new "urban night was not an extension of day; it was a liminal new world in which conflicting moral values mingled uneasily [and] … heightened the excitement of the nighttime street."[1] Most people still did not have electricity at home, and the brilliant, pulsating lights of downtown were an alluring contrast to domestic space, even if reformers thought it garish. Houses and apartments began to seem dim, especially if invaded by light from the street. Reformers also objected to the skyscraper because it blocked sunlight, created powerful winds, and disrupted the human scale of the city. Such views prevailed in Boston, where few tall buildings were erected before 1920. Reformers disliked large (especially flashing) advertising signs. They believed that "the individual has no more right to offend the public's eye with flaunting self-assertiveness than to offend its ear with crashing sounds or the nostrils with unpleasant odors."[2] They were pleased that there existed a "British Society for Checking the Abuses of Public

Advertising," and that there were strict controls on posters in Paris and many other European cities.[3] They applauded when Berlin issued "stringent regulations as to the number, size and height" of electric signs, and further restricted "flash signs" on the grounds that they were injurious to the eyes and a dangerous distraction to traffic.[4] And they objected to the fragmentation and incoherence of US public space, and hoped through planning to recover a sense of cohesion, order, and civility.

The excesses of advertising provoked much complaint. Landscape architect Frederick Law Olmsted had been one of the designers of the Chicago Exposition, which had not allowed any outdoor advertising. He was dismayed, however, by the commercial displays inside the exhibit halls.[5] Olmsted did not think legal sanctions could entirely prevent outdoor signs, as landowners had the right to operate and advertise a business on their property, and it would be difficult to infringe that right. But he wanted to prohibit advertisements near government buildings, parks, and boulevards, or on sites where an advertiser did not have a business. Olmsted thought it possible to educate popular sentiment so that advertising excesses would prove bad for business.[6] *Current Literature* took the same position when New York City's aldermen debated "the possibility of restricting the size of advertising, billboards, and sky signs."[7] The *Los Angeles Times* called electric advertising "eyesores" that were "distracting and chaotic disfigurements" of the city. It complained that "just at the moment when the demand for municipal art for more beautiful cities ... has become almost universal, here descends upon us a plague of outdoor advertising—sky signs and billboards—so aggravated, so acute, as to bid fair to nullify the great part of the benefits already attained in making 'A City Beautiful.'" Even Boston's colonial landmarks "had been disfigured" and become a "hub of panacea advertisers." This was not only a problem in New York, Philadelphia, and Chicago but also in "such resorts as Salt Lake, Tampa, Saratoga, etc." To combat this visual blight "an active crusade" had begun. The paper denied the advertisers' claim that the public "rather delights in feasting its eyes on the vociferous proclamations of the merits of rival soaps, nostrums, whiskies and cathartics." It might be that "the rough school of commercialism" had become insensitive to "beauty in public places," but the newspaper declared that most people found "these signs, with their violence to the sense of public order, [and] their disagreement with the character of the buildings that they disfigure ... repugnant."[8] A letter to the *New York Times* asserted that the city was "decades behind Paris, Berlin, London and every other civilized capital." The writer believed that for most people, the "principle interest in the new electric curiosities" was "the ingenuity of the electrician," while "the thing advertised is of little consequence."[9]

These pressures came to a head in 1914 when New York City passed a billboard ordinance that permitted new or existing electric signs only if they met certain provisions. On the roof of a building, signs could be no more than seventy-five feet high, and the building had to be fireproof. Signs had to be set back six feet from the edge of the roof, and needed to provide seven feet of headroom underneath in case firefighters needed access. Aside from these limitations, the content and appearance of the advertisements were not controlled.[10] The battle had resulted in a compromise that kept some flagrant signs off some streets, notably New York's Fifth Avenue, but otherwise left advertisers free to continue as before. Similarly, attacks on skyscrapers led to restrictions on bulk but not height.[11] Skyscrapers could not rise straight up from the sidewalk but instead had to use setbacks, and towers were restricted to a certain percentage of their lot. Such rules ensured more sunlight and fresh air at the street level, but did not impose an aesthetic.

Reformers were able to curb some excesses of commercialism, and they improved the appearance of some streets, but the visions of the great expositions were not realized wholesale. This may have been the best result, for as Jane Jacobs once observed, "The remarkable intricacy and liveliness of downtown can never be created by the abstract logic of a few men. Downtown has had the capability of providing something for everybody only because it has been created by everybody."[12] Had the City Beautiful movement been fully realized, it might have become the city sterile, as was the case with many grand designs of city planning in the 1960s.

Exposition fairgrounds themselves were invariably torn down. A journalist noted in 1901 that the "only remaining vestiges" of Chicago's White City were the abandoned replica of Columbus's *Pinta* ship and, nearby, "a marsh from which rise the charred ends of piles." He wished a similar fate did not await the Buffalo Exposition, whose buildings might have become a model community. Every exposition demonstrated that cities could be more harmonious and attractive, but the evidence was ephemeral. Expositions quickly passed from idea to plan to realization to demolition, suggesting both the possibility of rapid transformation and difficulty of permanent change. The journalist asked, "Who believes that the people of the second half of our new century will be content to live in those abominations of desolation which we call our great cities—brick and mortar piled higgledy-piggledy, glaringly vulgar, stupidly offensive, insolently trespassing on the right to sunshine and fresh air, conglomerate result of a competitive individualism which takes no regard for the rights of one's neighbor?"[13]

Many Americans of the Progressive era believed that a better environment improved a citizen's character and forged a stronger sense of community.[14] Before the Chicago fair, Edward Bellamy's 1888 utopian novel *Looking Backward* had described an architecturally harmonious city in the year 2000. The book sold millions of copies and inspired a national society dedicated to realizing its vision.[15] More than 160 utopian works appeared during the following twelve years, and electrical innovations were one of their three most common themes.[16] They foresaw a world where smoke, disorder, noise, and class strife were replaced by clean, efficient, beautifully landscaped cities. Hard labor and long working hours disappeared in a society based on science and efficiency. Readers with such dreams glimpsed utopia again in the era's monumental expositions.

Nor was the sense of an attainable utopia limited to fairs and fiction. During the nineteenth century, both sanitary reform and park building had successfully shown that conscious designs could be imposed on cities, making them cleaner, healthier, and more pleasant. Such efforts presaged the development of city planning, which emerged alongside the City Beautiful movement between 1897 and 1905. Railroads constructed magnificent gateway stations, notably Washington's Union Station completed in 1907. Simultaneously, there was an "outburst of great civic schemes," including designs for Philadelphia, Saint Louis, Kansas City, Saint Paul, San Francisco, Cleveland, and Washington.[17]

Many visiting the new Panama Canal Zone saw it as a utopia embodying these ideas.[18] Journalists came primarily to see the enormous excavations, electric-powered locks, and electrified railroad built to haul ships through the canal, which would be operated around the clock under powerful lights. Authors such as Ray Stannard Baker were further impressed by the fair wages, social order, and technological modernity of "the Zone." The artist Joseph Pennell thought the "Zone is the best governed section of the United States."[19] Its sixty-two thousand inhabitants made 90 percent of their purchases at inexpensive commissary stores reminiscent of the emporiums described in Bellamy's utopia.[20] If the ideal United States seemed to be emerging in Panama, however, such views ignored the military's top-down control as well as the second-class status of West Indians brought there to build the canal. But white Americans thought the Zone fused advanced technology and tropical nature into a model progressive society. It was lavishly celebrated at the San Francisco Panama-Pacific Exposition of 1915, where one of the popular exhibits was a huge scale-model canal. Seated visitors were transported through fifteen locations around its edges, while listening to a recorded description of its wonders. The experience was much like viewing the Panama Canal Zone from an airplane.[21] The reengineering of nature was not unique to Panama. On a smaller scale, similar transformations had been championed by Olmsted, precursor of the City Beautiful movement and the leading US landscape architect for half a century until his death in 1903. (His son held a post at Harvard and brought landscaping traditions into the new field of urban planning.) When Olmsted senior created New York's Central Park, he moved thousands of tons of rock and dirt, and entirely reshaped the site so that it accorded with the aesthetics of the picturesque and beautiful. He landscaped many urban parks as well as Niagara Falls, Yosemite Valley, the Boston Fens, and Mount Royal in Montreal. Olmsted was already a national figure in 1866 when he lectured in San Francisco and drew up plans for its development. He helped Americans imagine more perfect cities, and inspired many who planned exposition grounds, parks, and new urban districts.

Public opinion seemed to be evolving higher standards, in part due to municipal art associations concerned with planning and urban aesthetics.[22] New York City established a Municipal Art Society in 1892, and there were similar groups in Cleveland, Chicago, Baltimore, and Cincinnati.[23] Nationally, the City Beautiful movement reached its peak between 1900 and 1910.[24] More than seventy-five civic improvement associations formed in all parts of the country, inspired in part by European cities that reined in the excesses of advertising. Berlin, for example, established "a system of neat pillars" on street corners, where attractive posters were permitted. The city retained ownership of these pillars and earned a tidy income from their use. Yet Berlin did permit "multicolored, changing, electrically illuminated sings on the tops and sides of businesses," and these made the city at night "dazzlingly brilliant."[25]

Charles Mulford Robinson drew many European comparisons in his widely read *The Improvement of Towns and Cities; or, the Practical Basis of Civic Aesthetics*.[26] He advocated planning commissions staffed by representatives from engineering, architecture, landscaping, and art. Robinson combined these interests as a lecturer and consultant. Commissions in his native New York State included Rochester, Binghamton, and Jamestown, but he worked all over the country, including Detroit, Pittsburgh, Columbus, Denver, Oakland, Sacramento, Santa Barbara, Los Angeles, and Honolulu. He became a professor of civic design at the University of Illinois, imparting his views to another generation of planners. Burnham was another influential figure. He planned the Columbian Exposition and was an active consultant afterward, including a detailed plan for the future development of Chicago.[27] In 1904, he was invited to San Francisco to develop a city plan. Burnham extracted principles of design from Paris, Berlin, Vienna, and London, and suggested ways that San Francisco could learn from them "to alleviate the faults of the gridiron pattern" with curving streets on the steepest hills and traffic circles inspired by Paris.[28] Like Robinson, he championed horizontal urban designs with a unifying aesthetic based on the Beaux-Arts tradition.

Robinson's *The Improvement of Towns and Cities* saw electrification as an essential part of a coming transformation. On its first pages he explained, "An electrical age may relieve the city builder from the unhappy sacrifice of the water view in an industrial community." Water-driven mills needed to be close to rivers, but electric power could "drive profitably a thousand manufactories of which not one would be in sight from the river." After electrification, the shoreline could be recovered for parks, housing, and civic space. Robinson favored moving factories and working-class housing toward the periphery, so that old factory districts and the city center could be redeveloped. This arrangement would put workers closer to the countryside, with easy access to city centers using electric trolley lines.[29] He saw the street not as a mere transportation artery but instead as a place of sociability that should be wide enough to contain lines of trees and accommodate a public *salon des fêtes*. With electric lighting, it could be "as bright and gay at night when work is done as it is convenient by day." Planned lighting was essential to a city that encouraged the mingling of social classes in public space. Robinson advocated high levels of illumination. "Garishness has resulted now and then, but it has been by private extravagance supplementing, for its own ends, the public lighting. As far as the city's street lights are concerned, a *ville lumière* is still the goal."[30] It was not to be achieved by allowing free play to commercial forces. Rather, cities needed tight regulation of advertising. He praised Glasgow because it had sacrificed £4,000 a year in revenue so "that the municipal trams should not be disfigured by advertising" and was delighted that other British cities had followed this example, including Liverpool, Hull, and Sheffield. He was pleased to report that in Glasgow, "flashing electric signs have been prohibited in various places," and in London, three hundred architects signed a petition in 1899 seeking "to repress the evils of monstrous letter and illuminated advertising."[31] Robinson also praised Chicago for passing "an ordinance which limited the height of buildings to 130 feet," and Boston for resisting skyscraper construction and legislating against it in Copley Square.[32] These were short-lived victories that signaled the City Beautiful movement's preference for a horizontal city open to the sun. Yet powerful business interests pushed for more advertising, particularly members of the National Electric Light Association. The association's conventions focused on how to increase demand for electric advertising. One member remarked in 1908 that even in a community of only twenty thousand people, a utility could make almost a dollar per inhabitant per year on sales of electricity for signs.[33]

The City Beautiful movement worked against such competitive individualism and favored coordinated lighting. Robinson acknowledged the Columbian Exposition as an early inspiration. "The fair gave tangible shape to a desire" along with impetus.[34] There was also some synergy between the City Beautiful movement and electrical engineers, many of whom were progressives.[35] Americans then saw engineers not as apolitical technicians but rather as heroes who brought the forces of nature under control and made them serve humankind. Engineers seemed to a writer in the *Atlantic Monthly* the poets of the machine age.[36] They built dams, opened the west to irrigation, and conquered space and time, spearheading progress. Many lighting engineers valued city planning and believed new technologies would improve the urban environment. In 1906, several hundred of them met in New York City and formed the American Society of Illuminating Engineers. Their annual meetings and journal advocated better lighting as a civic improvement, aid to education, and embellishment of the arts as well as sound business investment.

In a speech in Cleveland at General Electric's National Lamp Works, chief engineer S. E. Doane underlined the shared interests of engineers and reformers. His large audience included Brush, who had pioneered arc lighting, along with members of the Illuminating Engineering Society, American Institute of Engineers, and Cleveland Advertising Club. Doane argued it was an engineer's civic duty to beautify a city. He praised European rules against indiscriminate advertising, advocated the elimination of many unsightly poles by attaching wires to the sides of buildings, and complained of the needless multiplication of poles that together with "letterboxes, waste paper boxes, street sign posts" and other objects provided "a natural accumulation place for dirt and rubbish of all kinds." Doane also praised Cleveland's new White Way, whose ornamental poles drew artistic inspiration from installations in Berlin, Paris, and Vienna, but had superior reflectors and lighting designed to spread the light more evenly.[37]

Progressive engineers like Doane believed intensified lighting combated crime, reduced accidents, and drew citizens into public space at night. Alongside the City Beautiful reformers, they lobbied for ornamental light standards that harmonized with local architecture (see figure 7.1).[38] Cities were keen to have lighting that was not only brighter but more artistic too. Saint Louis had an architect design new standards, installed in 1909. That same year, Boston sent Louis Bell to twenty European cities to study their lighting systems.[39] To serve to such customers, General Electric hired and trained "a new kind of salesman, the street lighting specialist," who not only understood the technical side. "He had to have much of the artist, something of the architect, and very much

7.1 Ornamental Streetlight, Wilshire Boulevard, Los Angeles, 1914
Source: General Electric trade catalog, 1915, Smithsonian Institution, Washington, DC

of the engineer about him."[40] Such salesmen sold new white way systems with distinctive designs like that in Cleveland. San Bernardino, California, had "giant arrowheads topping its posts," while nearby Riverside opted for standards "with a mission flavor," topped by a cross.[41] These new systems had opalescent glass globes that reduced glare and diffused the light more evenly, including the upper portions of buildings. Such white way installations spread beyond central business districts into other major streets, but did not always cost more to operate, because utility power stations were achieving economies of scale.[42] They extracted more electricity from each ton of coal, and more efficient lighting was available, notably the gas-filled tungsten street lamp that appeared in 1914.[43] To satisfy the demand for aesthetically pleasing lighting systems, General Electric hired a professional artist, Joseph W. Gosling, to "design artistic standards and artistic fixtures." He also prepared for customers "a vivid picture, in colors, exactly reproducing a specific street" as it would look at night with a new lighting installation. Viewing such images helped city committees when selecting a system.[44] Advertising for white ways stressed not just the improved lighting but also benefits such as crime deterrence, reduction in accidents, discouragement of littering, and stimulation of civic pride.

The white way sales campaigns appealed to progressive movements that wanted brighter, safer streets and evening access to parks as retreats from noise and congestion, where all social classes shared common enjoyments. In a number of communities, the City Beautiful movement was strong enough to impose new urban designs, notably in Harrisburg, Pennsylvania, and the trans-Mississippi west, such as in Kansas City, Denver, Dallas, and Seattle. Ideal urban landscapes, enhanced with spectacular lighting, were also erected at expositions in aspiring western cities, including Portland's 1905 Lewis and Clark Exposition, Seattle's 1909 Alaska, Yukon, Pacific Exposition, and San Diego's 1915–1916 Panama-California Exposition. But the most important events in disseminating electrical advances were the great expositions held in Saint Louis in 1904 and San Francisco in 1915.

The Saint Louis Exposition marked the centennial of Thomas Jefferson's Louisiana Purchase from Napoléon in 1803. At a stroke, the Louisiana Purchase had doubled the size of the United States and opened the Mississippi Valley to settlement. Saint Louis saw Chicago as its only rival for supremacy in the heart of the continent. With a population of half a million, Saint Louis was the nation's fourth-largest city. Its fair was almost twice as large as Chicago's a decade earlier, including a stadium where it staged the first Olympic Games held in North America. An enormous Column of Progress towered over the central square, and its promoters "placed St. Louis at the pinnacle of progress."[45] Advanced technologies, particularly electrification, demonstrated cultural superiority. A world's fair was to be not merely entertaining but also educational, and, as in Omaha and Buffalo, "ethnography exhibits" presented "native villages" as a contrast to US civilization. A large US government compound displayed a thousand Filipinos, including Igorote tribespeople, as part of an evolutionary presentation of "the four culture grades of savagery, barbarism, civilization, and enlightenment." The United States and the exposition exemplified the highest stage, while anthropologist W. J. McGee presented Pygmies, Patagonians, and Kwakiutls from Vancouver Island as less developed cultures.[46] The world's fairs disseminated a model of progress in which science discovered new knowledge, technologists applied it to everyday use, and industrialists spread the material benefits to the general population, in a process of continual uplift. As Michael Adas established in *Machines as the Measure of Men*, cultural differences during the nineteenth century were attributed less to inherent racial traits than to differences in science and technology.[47] In this context, lighting was not merely a convenience but also proof that scientific research led directly to cultural uplift.

The City Beautiful movement was part of this vision of evolutionary progress, both on the fairgrounds and in Saint Louis's plans for civic improvements.[48] After the idea for a comprehensive model city at the Buffalo fair was dropped due to insufficient funds, the American League of Civic Improvement convinced the Saint Louis Exposition to include such a model city. The architect Albert Kelsey worked in consultation with Robinson to develop a ten-acre exhibit. There were to be no overhead wires, no chimneys, and no billboards. A train station would serve as the entrance, and disembarking visitors would see inspiring full-scale reconstructions of sections of famous boulevards in Turin, Vienna, Paris, and London before coming to an exemplary US city center.[49] This grand plan was not built because it cost too much. Instead, a modest exhibit of four city blocks was "laid out as a curved street, spreading out into a square at the center," with buildings "appropriate for a town of ten thousand people."[50] At one end was an intramural rail station, donated by Atlanta. The buildings were "of imitation red brick with white trimmings and pillars," in an updated Georgian style. There was a city hall, model playground erected by New York City, school building donated by Missouri, auditorium and model worker's home donated by Dayton, and civic museum built by Minneapolis–Saint Paul. More than twenty cities collaborated, and the exhibit included photographic displays of projects from around the country.

Less money was invested in this City Beautiful exhibit than in the enormous pentagon of the Electricity Building, which covered three hundred thousand square feet, four times larger than Buffalo's electrical exhibit. Every imaginable device seemed to be there. At one popular exhibit, visitors compared how different kinds of lights illuminated colored and patterned fabrics. Enclosed arc lights gave a bluish white light; the Bremer arc light was more yellow; the magnetite arc lamp provided an extremely white light; the Nernst lamp gave a yellowish white light; and the mercury vapor lamp imparted a greenish tinge.[51] Another popular electrical exhibit was an incandescent light factory that produced two thousand lamps each day.[52] Outside, the new mercury vapor tube lights gave a green hue to the amusement area. The Nernst light was prominent in the Fine Arts Building, chosen because of its light spectrum and steadiness, as it was little affected by voltage fluctuations.

As in Omaha and Buffalo, evening crowds witnessed a general illumination that showcased a new technique. Rather than turning on rows or banks of lights in stages, the light first came on at a single point and raced out from there in all directions as fast as the eye could follow it along the rim line of the buildings. It moved like lightning, and in its wake hundreds of thousands of bulbs blinked on (see figure 7.2). A popular song suggested the central place of lighting: "Meet me in St. Louie, Louie, meet me at the Fair / Don't tell me the lights are shining any place but there."[53] *The Book of the Fair* proposed that the ideal way to see the lighting was to take the streetcar to the fair after dark, in order to view it first as "a mighty bouquet of light blossoming out of the darkness." As one drew closer, "a huge star breaks out, made of many

7.2 Machinery Hall, Saint Louis Exposition
Source: Hall of History, Schenectady, NY

lights," and the visitor marveled at "the ivory tinted exteriors of the huge buildings, glowing in the light of thousands of lamps." The scene seemed as bright as day, but with a new palette of colors. The "flower beds took on fantastic hues," and the scene continually altered as "the lights change and change in bewildering variety."[54] Buffalo's Pan-American Exposition had staged a single, defamiliarizing moment in a ceremony at dusk. In contrast, the Saint Louis Exposition, like the giant advertising signs in Times Square, continually altered its appearance, but its alternations were part of a coherent design. Henry Adams looked at the spectacle with "iniquitous rapture," declaring, "The world had never witnessed so marvelous a phantasm," not least because it had been produced by "a third-rate town of half-a-million people." He "wandered among long lines of white palaces, exquisitely lighted by thousands on thousands of electric candles, soft, rich, shadowy, palpable in their sensuous depths."[55] The lighting at Saint Louis had advanced far beyond what Adams had seen in Paris in 1900. But the Louisiana Purchase Exposition was the last major fair to rely on pointillism using small bulbs. Spectacular lighting would soon stop treating buildings as scaffolding to hold thousands of individual lights.

The tungsten lamp invented in 1904 made it "possible to make concentrated light-sources" so powerful that "whole buildings and monuments could be flooded with light" from projectors hidden on rooftops, in trees, or at ground level.[56] Such projectors illuminated the Singer Tower, the Woolworth Building, and much of the Hudson-Fulton Celebration of 1909. J. P. Morgan and Andrew Carnegie headed the celebration's influential board, advised by General Electric lighting engineers. The scale of the event was unprecedented, as it was staged not only in New York City but also included the harbor, and stretched 150 miles up the Hudson River to Albany and Troy. The planners knew the exposition tradition in detail and were also inspired by "the illumination of the Champs Elysées in Paris at the time of the automobile parade in 1907."[57] The board initially apportioned $60,000 for lighting, and New York City added $65,000 more.[58] As preparations continued, lighting demanded additional sums.[59] The electrical utilities charged only half the normal price for electricity and spent $50,000 of their own to enhance the illuminations.[60] On the six-mile line of the parade march from 110th Street to Washington Square, iron poles were erected and incandescent lights were hung fifteen inches apart on both sides of the street. This alone required 25,500 lights. In the middle of the route was "the Court of Honor, with its glistening columns and its sparkling canopy" of overhead lights. This court was defined by "36 detached columns arranged in a double colonnade, 18 on each side of Fifth Avenue, along the curb lines from 40th street to 42nd street." Sixty feet high and designed to "harmonize with the columns of the New York Public Library, each was surmounted by a gilded sphere." The Court of Honor was further defined by stringing, "from capital to capital—along the curb lines, across the avenue, and diagonally—... festoons of laurel and smilax, intermingled with electric lights."[61]

These displays represented but a fraction of the overall illumination. Brooklyn Bridge was bedecked with 13,000 incandescent lights, outlining its architectural features, and another 36,000 lights were used on the Manhattan, Williamsburg, and Queensborough Bridges. The "great bridges appeared suspended in midair like vast festoons of sparkling gems, supported at their ends by pillars of lights."[62] Every public structure of importance was limned in incandescent lights, including the Washington

Square Arch, New York City Hall, and the Soldiers' and Sailors' Monument (see figure 7.3). Powerful searchlights were trained on Grant's Tomb and the Statue of Liberty, so they leaped out of the surrounding darkness. These were only the official illuminations. In addition, there were "the usual brilliant illuminations of the city, and elaborate private illuminations by the owners of large office buildings, stores, and dwelling houses." The Plaza Hotel was decorated with lights from sidewalk to roof, and the Waldorf Astoria Hotel was decked out in chains of lights that formed a giant pyramid. The many visiting ships in the harbor were also outlined in light.[63] These efforts combined "to convert the city into a veritable City of Light."[64]

7.3 Washington Square, Hudson-Fulton Celebration, New York, 1909
Source: Irma and Paul Milstein Division of US History, Local History, and Genealogy, New York Public Library

In contrast to the competitive jumble of signage in Times Square, the lighting of the Hudson-Fulton Celebration was carefully designed to achieve large, dramatic effects. The planners photographed "all the buildings, monuments and bridges which it was proposed to decorate by illumination … and upon these photographs were painted the outlines of the electric lights." These were reviewed by an "Illuminations Committee" and submitted to a "Committee on Decorations," which then harmonized separate initiatives. "When possible, designs were also submitted to the architects of the various structures" to improve the visual reinterpretation of their work with suggested refinements.[65] Furthermore, it was decided to place beacon fires on the highest places from Staten Island for 150 miles along the Hudson River to the "head of navigation" near Troy, bringing the second week to a brilliant end.[66] The illuminations in New York harbor used a new technology called "fireless fireworks." The lighting equipment filled the equivalent of an entire block along the Hudson at 155th Street. The lights used as a medium smoke bombs and steam generated by a large boiler, vented from a row of stacks along the shore. This fogged atmosphere was painted with light, using searchlights, scintillators, and filters. Rainbows soared over the Hudson, and the sky was filled with a giant peacock's tail. Unlike conventional fireworks, these visions did not fall away but rather kept evolving and changing. The electric scintillator's operator could choose between nozzles for pinwheels, a fan, a snake, a plume, a column, or a sunburst as well as the spectacular Niagara nozzle. These effects could be enhanced using image projectors, prismatic reflectors, flashers, and filters to produce uninterrupted, kaleidoscopic effects, culminating in a powerful aurora (see figure 7.4). Newspapers reported that ecstatic crowds lined both sides of the river and watched until after midnight.[67]

The techniques used at the Hudson-Fulton Celebration were developed further at the 1915 San Francisco Exposition, which celebrated both the completion of the Panama Canal and the rebirth of San Francisco after its earthquake and fire. Like the fairs in Chicago, Omaha, and Saint Louis, the Panama-Pacific Exposition's architecture was a derivative of Beaux-Arts classicism, albeit with Spanish colonial ornamentation. It was a horizontal landscape, with the exception of the forty-three-story "Tower of Jewells," before which stood a large "Fountain of

7.4 Hudson-Fulton Celebration, Aurora, 1909
Source: Hall of History, Schenectady, NY

Energy." The spectacular lighting was again the work of Ryan. He and the theater designer Jules Guerin, the fair's chief of color, rejected white exposition buildings because recent research showed that the ancient Greeks had painted their statues and colored their temple walls.[68] Guerin developed a comprehensive color scheme for the fair in which the "first tonal value was the travertine, and on this travertine the other colors were applied; always having in mind the strong light of California, and keeping colors well toned down." Ryan determined his lighting would not obstruct or obscure Guerin's colors but instead enhance them. He saw that "previous exposition buildings have, in the main, been used as a background on which to display lamps. The art of outlining, notably the effects obtained at the Pan-American Exposition at Buffalo, could probably not be surpassed." Such lighting had "been extended to amusement parks throughout the world" and become "commonplace."Yet they suppressed "the architecture which becomes secondary, and it is practically impossible to obtain a variety of effects, so that the Exposition from every point of view presents more or less similarity." In addition, "the glare from so many exposed sources particularly when assembled on light colored buildings causes eye strain."[69]

Rather than limn the ornamentation of buildings, Ryan used "masked lighting diffused upon softly illuminated facades," and emphasized "strongly illuminated towers, and minarets in beautiful color tones."[70] The hidden lights provided "an even glow seeming to come less from a specific source than emanating from the walls themselves" (see figure 7.5).[71] To avoid flattening the structures' appearance, additional lights were employed to cast reddish "shadows" that added a sense of depth to the picture. The windows had a new function at night, as a faint "warm orange light" emanated from them, suggesting animated life within. Ryan realized that uniformity could be deadly, and supplied a "contrast to the soft illumination of the palaces" through the lighting of the "amusement section with all the glare of the bizarre," which also gave "the visitor an opportunity to contrast the light of the present with the illumination of the future."[72] The lighting plan erased the coming of night, as the lighting gradually came on to replace natural light. Ryan wanted his light to be "so perfectly and unobtrusively distributed that nine-tenths of the people will not know" that night had fallen.[73]

The fair's lighting did not preserve the buildings' daytime appearance, particularly the central tower. Embedded in its surfaces were 102,000 "Novagems," cut Austrian glass in five different colors that refracted and reflected both natural and electrical illumination. The Novagems were hung on wires so that they could move in the wind and enliven the scene with shimmering refractions. The buildings had an otherworldly iridescence, bringing visitors back often. The lighting effects were predominantly green for Saint Patrick's Day and simulated a great conflagration on the anniversary of the San Francisco fire.[74] "Concealed ruby lights, and pans of red fire behind the colonnades on the different galleries seemed

7.5 Palace of Fine Arts, Panama-Pacific Exposition, San Francisco, 1915
Source: Prints and Photographs Division, Library of Congress, Washington, DC

to turn the whole gigantic structure into a pyramid of incandescent metal, glowing toward white heat and about to melt. From the great vaulted base to the top of the sphere, it had the unstable effulgence of a charge in a furnace, and yet it did not melt, however much you expected it to, but stood and burned like some sentient thing doomed to eternal torment."[75]

Luckiesh summarized Ryan's work as "a great variety of direct, masked, concealed, and projected effects … blended harmoniously with one another and with the decorative and architectural details of the structures." The Panama-Pacific Exposition "was a silent but pulsating display of grandeur dwarfing into insignificance the aurora borealis in its most resplendent moments."[76] The exposition gradually turned on its lights at dusk, adding special effects as the darkness increased. "Tall Venetian masts topped with shields and banners directed light from powerful magnetite arcs at the walls of the palaces, bathing them with a soft, shadowless radiance. Perfect reflections were thus assured in still pools in the courtyards. Searchlights on the roofs of the palaces and the towers raked the sky and spotlighted heroic sculpture on the skyline, casting their silhouettes through the fog." Each courtyard had distinctive lighting: green waters were in the Court of the Seasons, the Court of the Universe celebrated the rising and setting sun with whitewater fountains, and the Court of the Ages used crimson special effects, including what seemed to be serpents rising in the mist. The glass dome of the Palace of Horticulture became the translucent screen for an "electric kaleidoscopic" that projected moving patterns, including what appeared to be comets and planets streaking along its surface.[77]

A scintillator was positioned to shine through the fog banks that usually rolled in at night. A battery of forty-eight powerful searchlights was placed in the harbor, and three nights a week, at 7:45 p.m., it was "manned by a company of marines" who "executed precise drills … weaving artificial auroras in the fog or, on clear nights, animating clouds of steam lofted by a stationary locomotive."[78] They could change the appearance of each light using filters and colored screens.[79] In one of the most impressive special effects, shortly after sundown the sun seemed to be rising in the West, followed by an artificial aurora borealis. The fireless fireworks provided a spectacular crown of light over the grounds, with scintillator effects whose names suggest their appearance: Scotch Plaids,

Fairy Feathers, Sun-Burst, Fighting Serpents, Chromatic Wheels, Plumes of Paradise, and Devil's Fan. Ryan even brought back gas "in street lighting in the foreign and state sections" as well as "gas flambeaux for special effects."[80] (Indeed, gas enjoyed something of a renaissance at the fair, as the Welsbach Company received the highest award for a bungalow displaying various gas mantles.)[81] The stunning visual effects were the talk of the fair. The popular poet Edward Markham raved about the illuminations, calling them "the greatest revelation of beauty that was ever seen on the earth. … [They] will give the world a new standard of art, and a joy in loveliness never before reached."[82] As Edward Graham Daves Rossell has observed, at the San Francisco Exposition, lighting engineers "learned to paint with light" and the scintillator displays were "an architecture of its own—actively shaping the space around it," as was also true of cities adopting white ways (see figure 7.6).[83]

Expositions presented each intensification of lighting as an improvement, and Westinghouse and General Electric sold these innovations as part of the white ways that made cities more impressive and attractive. A year after the Panama-Pacific Exposition closed, some of its equipment had been repurposed to create an intensive white way, San Francisco's "path of gold." To celebrate, the city held an illumination and parade on Market Street from the Ferry Building to the city center. There were "thousands of marchers, all in costume, and each carrying some form of light," interspersed with floats running on the streetcar line. The parade represented the history of illumination. Cave dwellers carried pine-knot torches, Assyrians marched with improved torches and lamps, the ancient Greeks had lamps, the Romans held candelabras, and on down to modern times, depicting the march of civilization toward ever-brighter cities, culminating in Market Street itself.[84] The new, monumental lamp standards shed eight hundred lumens per square foot, double the usual brightness in a business district. Buildings along the parade route also illuminated their facades, and prizes were awarded for the best displays.

Yet the Panama-Pacific Exposition did not set the pattern for US urban lighting, any more than tower lighting had. In everyday life, each business wanted uniformity less than it wanted to advertise itself and stand out. Businesses accepted the regularity of streetlights but resisted further imposition of grand designs. Instead, they embraced the aesthetics

7.6 Panama-Pacific Exposition, Night View
Source: Hall of History, Schenectady, NY

of Times Square. The riot of individualistic expression only looked harmonious when viewed from an airplane, skyscraper, or hilltop. Such vistas were by no means limited to New York City. A traveler from Oklahoma in 1915 extolled Pittsburgh as a "wonderful night scene" when viewed from the top of a ridge, asserting that "the city appeared a marvel of illumination; millions and millions of lights, like fireflies in the darkness, shining out from boulevards and by streets, outlining skyscrapers and huge electric signs." Along the rivers, "reflected lights and buildings cast mystic shadows upon the waters. … [T]he scene below was one of entrancing beauty, tranquil and yet so full of hidden life."[85] By 1915, one did not have to attend an exposition to see a stunning electrified vista.

The controversy over how to light US cities had resulted in a compromise between the City Beautiful movement and individualistic forces of commerce. National symbols were bathed in white light, including the White House, Washington Monument, city halls, and state capital buildings. Structures that had attained iconic status such as the Brooklyn Bridge received similar treatment. At national parks, a few natural sites had become national symbols, notably Old Faithful, and it too was spotlighted. Later, Mount Rushmore would be highlighted as well. The City Beautiful movement had successfully promoted tasteful lighting standards along major boulevards, and prevented some garish forms of lighting or at least restricted it somewhat. But commercial energies retained their focal points in theater districts, central squares, and amusement parks, where spectacular lighting effects had free play. Between these extremes stood the skyscrapers, which did not limit themselves to white lighting, as the Singer Building did in 1907, but instead added colors and special effects while seeking to retain some neoclassical dignity. The urban landscape that resulted was a hybrid form, neither the stately horizontal city of the great expositions nor the visual cacophony of Times Square. A compromise that admitted variety, verve, and occasional surprises, it lacked an intentional unifying style, but was interesting to pedestrians. If viewed from a skyscraper or airplane, it was impressive and unexpectedly attractive. The illuminated city expressed tensions between the Beaux-Arts tradition and US iconoclasm, between the horizontal city and vertical thrust of commerce, between an exuberant popular culture and reverence toward patriotic symbols.

8

Light as Political Spectacle

Illuminations had originated in the courts of Europe, where they served as a brilliant demonstration of cultural hegemony. In the democratic United States, people were long wary of monarchical pomp. The first presidents did not seek to celebrate themselves or stage magnificent displays. But this attitude changed over time. By the end of the nineteenth century, as in Renaissance Italy, Bourbon France, or Georgian Britain, spectacular lighting had become part of political ceremony. Americans used it to dignify a presidential inauguration, mark an anniversary, drum up support for war bonds, or celebrate military victory. Illuminations were held in New York to celebrate the end of the Civil War, in Boston on the centennial of the Battle of Bunker Hill, and in Philadelphia on the centennial of American independence in 1876 (see figure 8.1). Electricity was adopted in Washington, DC, to illuminate major buildings and enhance special events. In 1878, when arc lights were quite new, the architect of the US Congress already was "making experiments with the electric light for the purpose of substituting it for the present expensive system of lighting by gas."[1] Elaborate illuminations quickly became a fixture at presidential inaugurations.[2] The *Boston Globe* reported in 1881 that "the illuminations are general and some are on an elaborate scale," including the White House grounds, which "attracted much admiration. A large illuminated star on the north portico was the chief feature. Lines of Chinese lanterns were stretched from tree to tree on both sides of the approaches from the avenue, and the shrubbery all around the grounds was similarly illuminated."[3]

8.1 Centennial of American Independence, Philadelphia, 1876
Source: Prints and Photographs Division, Library of Congress

On Memorial Day in 1891, President Benjamin Harrison described his surprise and pleasure at a spectacular lighting of the American flag:

> Two years ago … as we were going out of the harbor of Newport, about midnight, on a dark night, some of the officers of the torpedo station had prepared for us a beautiful surprise. The flag at the depot station was unseen in the darkness of the night, when suddenly electric searchlights were turned on it, bathing it in a flood of light. All below the flag was hidden, and it seemed not to touch the earth, but to hang from the battlements of heaven. It was as if heaven was approving the human liberty and human equality typified by that flag.[4]

Dramatic lighting made the flag into a patriotic abstraction, but this display only faintly anticipated the stunning visual effects achieved the following year.

In 1892, to celebrate the four hundredth anniversary of the discovery of the New World, the US Congress allocated $5 million to New York City to hold a commemorative event. Columbus Day itself was relatively a recent discovery, first celebrated in New York in the 1860s, where the annual parade started in 1869. By the centennial of American independence, Columbus Day had been adopted in Philadelphia, Boston, Saint Louis, Cincinnati, New Orleans, San Francisco, and other cities. Many streets were named in Columbus's honor. Notably, in 1892, New York designated a corner of Central Park "Columbus Circle," erected a statue of the admiral there, and dedicated it during the six-day pageant.[5] The city spent $250,000 just on signs, tableaux, and special effects, and held a night parade.[6] Sanford White and Louis Tiffany prepared a pamphlet that explained to homeowners and businesses how they should decorate, using either cheesecloth or bunting. For lighting, they recommended "lampions placed in rows on window sills and cornices." Lampions were "generally used in Europe on festal occasions" and consisted of "small glasses containing an illuminating composition cast in." They burned "with a bright light from six to eight hours," and were considered an "effective means for celebration and campaign illuminations."

The lampions were available at many locations in "red, white, blue, and green."[7]

If these private decorations were traditional, the official installations were state-of-the-art. The Edison Electric headquarters was bathed in the blaze of arc lamps, which the following year would dominate Chicago's Columbian Exposition. Red, white, and blue bands of incandescent lights bedecked the New York Life Insurance Building, along with one of the first large electric signs—a huge electric portrait of Columbus, in white and gold bulbs. The city also inaugurated a system of streetlights as part of the festivities. Recently returned from a European trip, New York's commissioner of public works declared, "No street lighting in Paris or London excels these lamps for beauty and illumination."[8] Several miles of marchers, 65,000 strong, paraded down Fifth Avenue under these lights, passing 2 million spectators. The parade included a series of horse-drawn floats, each lighted using batteries on board. In addition, "wires were run to lamps that participants in the procession carried while walking alongside." Twelve trumpeters "announced the first float, which featured winged Fame flying over a globe announcing America to the world." There followed a historical progression, beginning with a float devoted to cave dwellers, and another where an Incan priest "was sacrificing a human victim," followed by people dressed as Aztec warriors and a wigwam float accompanied by "5,000 members of the Improved Order of Redmen."[9] One massive float was devoted to Columbus, and several others celebrated the revolution and founders. Another presented a replica of the Statue of Liberty. The illuminated parade was considered one of the largest and most arresting civic events ever held in the United States.[10] At a time when few US homes had electricity, it was especially stunning.[11] The event also promoted US nationalism as, for the first time, thousands of schoolchildren recited the newly invented "Pledge of Allegiance" to the flag. The *Youth's Companion Magazine* had introduced "the pledge" a few years before, but it only won national endorsement after 1892.[12]

The patriotism of Columbus Day proved a prelude to the wildly enthusiastic reception of Admiral Thomas Dewey on his triumphant return in 1899 from his destruction of the Spanish Navy and conquest of the Philippines. In 1805, London had celebrated Nelson's victory at

Trafalgar with the best illuminations then possible. Almost a century later, New York staged a reception using all the technologies of lighting and fireworks developed since that time. As a centerpiece for the celebration, the New York legislature funded a temporary triumphal arch in the Beaux-Arts style, erected in Madison Square. Like world's fair buildings, it was made of wood and staff, but looked like an imposing permanent structure 100 feet high and 80 feet wide. The monument was illuminated, as were the harbor, rivers, and Brooklyn Bridge (see figure 8.2).

8.2 Dewey Arch, 1899, New York
Source: Prints and Photographs Division, Library of Congress, Washington, DC

Before the admiral arrived in New York City, he and his crew were obliged to stop at the port's quarantine station, decorated in their honor with 1,000 red, white, and blue electric lights.[13] That night, they could see an enormous "Welcome Dewey" electric sign on the Brooklyn Bridge. Each letter was 9 yards high, and the letter *W* alone required 1,000 bulbs.[14] On September 29, the formal event began, as the admiral's ship led a great parade of warships, yachts, and other vessels up the Hudson to Grant's Tomb. The shoreline was lighted as far as the eye could see that evening. A "mass of river craft in red, white and blue" floated on the tide, and eight barges loaded with fireworks began to shoot them off when it was fully dark. "There were two elaborate set pieces, one a portrait of Dewey and another a picture of [his ship] the Olympia 1,000 feet square. These were greeted by deafening salutes from the steam whistles," and cheers from the crews and multitude on the shore. At the same time, "scores of searchlights played about the harbor or bathed the skyscrapers in the radiance of a sun at night." The *Baltimore Sun* reported, "New York appeared to be an enchanted city. The great buildings were bright with light and gorgeous color."[15] The "city was aflame" with fireworks that "burst with joy," and "flaming fingers of electric lights crossed each other in the sky."[16] As the *Philadelphia Ledger* described it, "In the evening a fairy-like scene was presented by the illuminated war-ships, and fireworks blazed in every quarter of New York."[17] As *Harper's Weekly* declared in a bit of doggerel: "Let each fulgent 'lectric light / Fulge until it dazzles sight. / For 'tis plain to any dunce, / Dewey can come back but once."[18]

The following day, "nearly every office building on Broadway and the downtown thoroughfares" was "decorated with flags, bunting, shields, and streamers" as well as extensive "adornment of private houses."[19] More than a million people cheered the admiral as he progressed from Grant's Tomb to his reviewing stand next to the Dewey Arch in Madison Square. And when the last of the fifty thousand marchers had paraded by, including some veterans of the Civil War, "searchlights from the tops of the buildings played their rays upon the arch."[20]

Lighting focused attention on successful individuals, whether as tribute to a military hero, an actor's name in lights on a theater marquee, Woolworth's brilliantly illuminated skyscraper, or a politician's campaign billboard. Anyone walking the streets in 1910 could discern, based on the lighting alone, the centers of social life, names of major corporations, and individuals who had achieved fame, fortune, and power. This tight connection between lighting and status changed during the twentieth century, as the city deconcentrated, commerce relocated to the suburban periphery, and new forms of communication emerged. Later, celebrities were prominent on radio, then on television, and more recently through the Internet. But during the energy transition from 1875 to 1915, lighting was a dominant social media.[21]

Political parties adopted spectacular lighting in their parades, campaigns, and conventions, and elected officials embraced it for their inaugurations and public appearances. In 1891, for example, New York's Democratic Party put a steam engine and dynamo on a horse-drawn truck that displayed a flashing illuminated sign. While being pulled through the streets, the sign spelled out a candidate's name letter by letter, using 220 bulbs. Despite a rainy night, it worked for four hours "with no protection from the downpour" and "created immense enthusiasm" wherever it appeared.[22] One common dramatic effect at political conventions was to throw on red, white, and blue lights in quick succession while playing a patriotic song, which always brought the crowd to its feet.[23]

The night after elections, crowds filled the streets, seeking news, and eagerly buying extra editions of newspapers and reading them by lamplight (see figure 8.3). By the early 1890s, newspapers in Boston, Chicago, and New York used searchlights to signal election results for a public that had been told how to "read" them based on either the colored filter used or direction of the beam.[24] By the middle of the decade, newspapers adopted electric displays to attract crowds to their offices during elections (see figure 8.4). In November 1896, New Yorkers thronged City Hall Park, where the *New York Journal* had erected a temporary tower. The tower's powerful projectors beamed results on canvas sheets attached to the newspaper's building.[25] By 1901, this system had been replaced by an electric sign that had 4,600 lamps and could spell out messages of 300 letters.[26] This was the predecessor to the "zipper" that later spelled out the headlines in lights on the side of the New York Times Building. Such projections and signs differed from advertising, because their messages constantly changed to keep up with the news. During dramatic events such as the Spanish-American War or a close election, they could attract

8.3 Election Scene, 1876, New York
Source: Miriam and Ira D. Wallach Division of Art, Prints, and Photographs, Photography Collection, New York Public Library

and hold a crowd for hours. At the turn of the century, spectacular light was useful to politicians and journalists, who wanted to attract the new urban crowds just as much as advertisers did.

Lighting at presidential inaugurations became increasingly spectacular during the early twentieth century. Each was touted as being more extravagant than the last. In 1909 when William Howard Taft was sworn in, one headline declared, "Inaugural Was Most Splendid in History." Along Pennsylvania Avenue were erected "Venetian masts" holding "gilded baskets bearing greenery and flowers and festooned with gaily colored streamers." At street intersections stood Doric columns framing the entrance to each block. Special lighting effects heightened the sense of occasion, and more than a hundred thousand people were awed by the fireworks.[27]

8.4 Election Eve Crowd, Newspaper Row, New York, 1908
Source: Miriam and Ira D. Wallach Division of Art, Prints, and Photographs, Photography Collection, New York Public Library

Four years later, the Democratic National Convention in Baltimore was spangled with lights both inside and out, when Woodrow Wilson won the nomination (see figure 8.5).

After Wilson won the 1912 election, at his inauguration fireworks and searchlights lighted up Washington, DC, and Pennsylvania Avenue "was transformed into a fairyland of light by hundreds of incandescent lamps spanning the street in graceful aches." The fireworks were coordinated with the firing of cannons, and the predominant colors were red, white, and blue. The crowning moment was a 2,000-square-foot lighting display that began as "a mammoth bouquet of roses" then changed into an American flag and finally became gigantic portraits of Wilson and Vice President Thomas R. Marshall. A typical newspaper story concluded, "Never, it is said, has so brilliant a display of fireworks or an illumination of such magnitude been attempted. Large searchlights threw their rays along the avenues and on public buildings, while at the monument groups the crashing of thousands of aerial bombs and the flaming light from large set pieces illuminated the sky."[28] The White House, lighted by

8.5 Democratic National Convention Hall, Baltimore, June 25, 1912
Source: Prints and Photographs Division, Library of Congress

gas in Lincoln's day, was permanently illuminated with electric lights (see figure 8.6). If Washington had been slow to adopt spectacular lighting, by 1916 it had caught up to New York and Chicago.

Attempts to reach and influence the crowd were not limited to politicians, advertising bureaus, and newspapers. The architect Claude Bragdon advocated "the combination of harmonic form, colored light, and choral song" in festivals that employed "projective ornament light fixtures." The first of these "Song and Light events" was held in 1914 in Rochester, New York, and then one in "New York's Central Park the following summer as well as festivals in 1917 and 1918." More than sixty thousand people attended the first Central Park event, which impressed Lewis Mumford, particularly the communal singing. The *New York Times* commented, "As the song rolled across the water from the absolutely

8.6 White House on a Winter Night, 1907
Source: Prints and Photographs Division, Library of Congress

invisible thousands, it had a strangely stirring effect, something that could not be duplicated in a thousand years of concert halls and opera houses."[29] Through active participation, the intention was to connect "a crowd of individuals into a unified polity expressing itself with one voice." A "community singing festival was geared toward replicating for mass society the political effect of classical drama: securing political consensus by uniting audience members in a shared catharsis capable of bridging ethnic and class divisions."[30] Bragdon considered ornamental lighting essential to these events and designed large screens, much like the transparencies long used in illuminations. These were backlighted by electricity and looked like stained glass windows. Together with Japanese lanterns ornamented in similar designs, "the psychological effect of the lighting on the people" was "to hypnotize them into a peace and quiet most favorable for the reception of the ever-changing message of the music."[31] Adding to the air of enchantment, Bragdon copied the Venetian practice of putting boats laden with paper lanterns on the lake, which separated the enormous crowd from the orchestra and eight hundred singers. After lighting set the scene and created a mood, it was also used dramatically. During quiet moments, light levels were low, but the "bright white lights were turned up at moments of greater intensity, spotlighting the chorus and sometimes illuminating the audience as they sang."[32] When the United States entered World War I, these techniques were adopted at rallies bidding farewell to departing troops. The Song and Light events had become part of the patriotic effort to rally the populace.

For electrical utilities, the war effort meant that many employees joined the army, even as demand for electricity dramatically increased. Replacing workers proved difficult. In Philadelphia alone, two hundred thousand additional people were hired for the war production of ships, rifles, and other weapons, and utilities found it hard to find and retain skilled laborers. The war also drove up prices for materials, and in winter 1917–1918 there was a coal shortage. Nationally, demand for electricity doubled between 1915 and 1919, and rationing of service for nondefense customers often became unavoidable.[33] Daylight saving time was made mandatory, which saved about 2.5 percent of the total coal production. Restricting electric advertising for much of the time saved only about 110,000 tons, but the dim downtowns sent a powerful message about the need for economy. The utilities successfully argued that signs should not be turned off all the time, because "the effect" on the public "would operate most effectively if the extinction of signs is intermittent rather than continuous, for much of the effect depends upon the contrast."[34] In New York, the public complained that the city did not seem normal when the Great White Way was dark, and eventually a compromise was worked out. The signs were relighted once they had been rewired to blazon patriotic messages and sell war bonds. There were still "lightless nights," however, to encourage the pubic to use less coal. It was estimated that if each of the 1.1 million families in New York saved one shovelful of coal a day, it would save almost a million tons annually.[35]

President Wilson often employed electrical signals and electrical lighting to participate in public events. The push-button activation of electrical devices gradually had become common after circa 1880. Grover Cleveland had opened the Chicago Columbian Exposition by pressing a button that activated the machinery and turned on the lights.[36] He was in Chicago in person, but in 1896 opened an event in Pittsburgh by pushing an electric button in Washington.[37] By the time Wilson entered the White House, there was frequently considerable distance from the button to whatever it started.[38] On April 24, 1913, at 7:30 p.m., while sitting in the White House, Wilson pressed a button that sent a signal to the engine room of the Woolworth Building, the world's tallest tower, and its "eighty thousand incandescent lights flashed on all at once" and were greeted by the cheers of a vast crowd of tens of thousands.[39] Later that year he pressed a button in the White House that triggered a telegraphic connection to Panama, where eight tons of dynamite exploded, blowing up a dike that was the last barrier to the waters that flowed into the newly completed Panama Canal.[40] And in November 1914, the president pressed a button that triggered a blast from a mortar gun to signal the opening of a new canal in the Port of Houston.[41] On January 24, 1915, he participated in the first transcontinental telephone call, speaking with the telephone's inventor, Alexander Graham Bell, and the assistant who had helped to invent the first telephone in 1876, Thomas Watson, who received the call in San Francisco.[42] Four weeks later, Wilson pressed a gold telegraph key in the White House that sent a wireless signal to San Francisco, putting into motion the machinery of the

Panama-Pacific Exposition and turning on its lights. From afar he also opened San Diego's exposition that year, setting off a fireworks display.[43] Wilson made extensive use of electrical technologies—telegraph, wireless, telephone, and electrical lines—in order to make his presence felt even when he was a continent away. Push-button participation required no travel and little time, and it was possible for him to signal support for even small, local celebrations, such as the centennial of Illinois's Fort Armstrong.[44] These media events were carefully prepared in advance to address a broad public.

Even as push buttons became a central part of presidential public relations, brilliant electric lighting was becoming intertwined with nationalism. On Independence Day, 1916, New York City banned fireworks and installed "special lighting for the evening" that totaled fifty-five million candlepower. The major illuminations were at New York City Hall and the Soldiers' and Sailors' Monument on Riverside Drive, but "practically every monument, park, square and public place in the city" was "made brilliant." Earlier in the day, Union Square was filled with "foreign singing societies" which belted out patriotic songs for several hours, while nearby booths provided information and encouragement to anyone wanting to take out naturalization papers. That evening, President Wilson delivered a short "silent speech" that was "transcribed in electric lights and flashed" to crowds at four New York locations: City Hall Park, Times Square, Columbus Circle, and the Stadium of City College. Wilson's message was sent as a rolling text, using the technology perfected for advertising signs as a form of broadcasting. For the time period, this was a sophisticated media event that simultaneously reached four crowds and could have reached many more. The electric sign itself materialized a patriotic metaphor, as Wilson concluded, "May the light of America grow brighter with each generation and blaze out pure and undefiled to all the world."[45] During the following years, the Wilson administration would use techniques of advertising and public relations to rally support for the war.

In summer 1916, "electric American flags" became popular. The Federal Sign System Company in Chicago advertised to utilities, "Boost America First Signs in your locality. Cash in on the spirit of patriotism which is engulfing the country."[46] A Toledo utility sold more than three hundred electric flag signs, and a Dayton utility ordered them in batches of fifty. In Athens, Georgia, the local utility successfully promoted sales of electric flags, including an especially large one that the city lighted at an evening ceremony with three thousand people attending. The mayor had proclaimed a holiday, and all the stores were decked out in red, white, and blue bunting.

Amid such patriotic fervor, public monuments began to seem too dark, and steps were taken to bring them literally back into the light. When the Statue of Liberty was erected, only the lamp she held aloft was illuminated. Its radiance was exaggerated on postcards, but the reality disappointed. As the rest of New York became more brilliant, the statue became "a speck of light more feeble than many surrounding shore lights." The statue was becoming obscure at the very moment that she was replacing Uncle Sam as the most popular national symbol.[47] The *New York World* raised funds in a subscription campaign, H. H. Magdsick was appointed as lighting engineer, and 225 searchlights were installed that bathed the statue in 20,000,000 candlepower.[48] President Wilson and the French ambassador came from Washington to inaugurate the new system, and in response the skyscrapers of the city turned on all their lights.[49] At 6:00 p.m. on December 2, 1916, Wilson pressed a button that set the lights ablaze, while "an illuminated aeroplane" passed overhead with "lighted letters three feet high" that spelled "Liberty." The scene was further "lighted by a 1,250,000,000 candle power Sperry searchlight, the most powerful in the world." Afterward, the president rode in a "procession of lighted automobiles" down Fifth Avenue, passing five thousand Boy Scouts and five thousand park playground children standing at solemn attention.[50] In the harbor, the brilliantly lighted statue became a vivid reference point for passing tugboats, ocean liners, and ferries as well as soldiers as they shipped out to Europe (see figure 8.7). In the iconography of World War I, notably in drives to sell war bonds, the Statue of Liberty held aloft an electrified torch of democracy.[51]

The electrical utilities, under the leadership of Western Electric, General Electric, and Westinghouse, used the lighting of the Statue of Liberty as the kickoff event for a national "electrical week" from December 2 to 9, 1916. They sought to reach every US city or town of more than ten thousand people, with "parades and pageantry, special illuminations,

8.7 Relighted Statue of Liberty, 1917
Source: Miriam and Ira D. Wallach Division of Art, Prints, and Photographs, Photography Collection, New York Public Library

electrical shows and demonstrations of what electricity has done in peace and in war." In Macon, Georgia, the city planned an electrical parade, demonstrations of new appliances, and special additional lighting of the main streets.[52] Larger cities, such as Philadelphia, had more extensive illuminations.[53] In Washington, Pennsylvania Avenue was "a blaze of illumination from electrically lighted display signs and specially installed flood lights, which swept the facades of many public buildings." Down the street came "a brilliant caravan of electric and motor propelled vehicles" that "filed through an aisle of thousands of spectators" in order "to tell them that this is electric week and that electricity has done more in modern achievement than any other scientific agency."[54] The many floats advertised electric stoves, vacuum cleaners, portable lamps, and other appliances. Such consumer displays would seem inappropriate once the nation entered the war, however. A few months later, at Wilson's second inaugural, Washington again was ablaze with illuminations. From the roof of the Capitol flowed a stream of light over the city. Giant searchlights were trained on every monument, such as "the Peace Monument" and many temporary decorations, including "columns and pylons twenty feet high surmounted by globes and eagles with outstretched wings" along with "huge urns, filled with blazing fire."[55]

After the United States entered the war, Wilson appointed George Creel to head a new agency that used all the techniques of advertising and public relations to unify the public behind the president. A journalist by profession, Creel thought that during the neutrality of the first war years, "the United States had been torn by a thousand diverse prejudices, with public opinion stunned and muddled by the pull and haul of Allied and German propaganda."[56] There was considerable isolationist sentiment, and some argued that it was a "rich man's war" that would save Wall Street loans. Some Irish Americans wanted Britain to lose, while many German Americans were ambivalent. Creel's mission was to rally support for the Allies and sell Wilsonian rhetoric about "making the world safe for democracy." He set up "twenty-one divisions devoted to domestic propaganda," including one on advertising that took a strong interest in outdoor advertising. He declared, "The printed word might not be read, people might choose not to attend meetings or to watch motion pictures, but the billboard was something that caught even the most indifferent eye."[57] Creel recruited the best illustrators in the country and members of the leading New York advertising agencies as well as major corporations. They not only did the work pro bono but also paid for rewiring signs, the posters, and advertising in newspapers.[58] The eighty-eight firms that made up the Outdoor Advertising Association "contributed more than $150,000 in advertising in support of war policies," as they promoted Liberty bonds, war savings stamps, and the Red Cross.[59] After years of being attacked by the Outdoor Art Association that "made war on billboards," these wartime contributions improved the public image of advertisers, demonstrated their patriotism, and silenced their critics.[60]

Close cooperation with Creel also had immediate benefits. After the Federal Fuel Administration decreed that the electric signs on Broadway be turned off to save coal, Creel successfully argued that they should be turned on again with new messages appropriate to the war effort. The largest electric sign of all, covering an entire block, was owned by Wrigley's chewing gum. On it, two giant peacocks faced each other at the center, with "their tails forming a feathery canopy" over Wrigley's name. "Flanking the text were six animated 'Spear men,' three on each side. … Brandishing spears, they comprised a drill team that went through a series of twelve calisthenics the populace quickly dubbed 'the daily dozen.' Flanking them were fountains spraying geysers of bubbling water."[61] Most of the sign remained the same for the war effort, but the text beneath the two peacocks was rewired to read, "Your patriotic duty—Buy a Liberty Bond."[62] Dramatic lighting was also used in parades that called on citizens to buy Liberty bonds. In May 1919, New York's Fifth Avenue was "ablaze with red illuminations, lighting the way of 5,000 marchers who opened the Red Feather campaign for the Victory Loan."[63] All the marchers wore a red feather, which indicated that they had bought at least two "Victory Notes." At the head of the parade was New York governor Alfred E. Smith and the assistant secretary of the treasury, Martin Vogel, and several military bands were interspersed among the civilian marchers.

For the common soldier on the battlefield, it was dangerous to light a fire or even a cigarette, as it pinpointed a person's face for a sniper. A soldier was safer in the dark. But electric light played an important role in the Great War, which used more industrial technologies than any previous conflict. Electric lights were essential inside otherwise-dark tanks and submarines. They marked the location of airstrips, and illuminated hospitals and military buildings. Searchlights crisscrossed no-man's-land, searched the sea for enemy ships, and raked the skies looking for enemy planes. Electric lights were also employed in signaling, and a new transatlantic radio system communicated with allies across the Atlantic.

Yet most important of all, on the long western front that stretched across France to the Swiss border, hundreds of miles of trenches gradually were wired into a comprehensive infrastructure that included signaling, lighting, telegraph lines, and telephones.[64] Before the war began, neither the British nor the French had organized electrical specialists into units, but they found it essential to do so, as the war proved not to be the nine-month offensive that they had expected but a long stalemate. To meet the demand for electricity, the Allies had to rely on a patchwork of small, isolated plants as well as some service from civilian power plants near the front lines. The US Army observed these difficulties and was better prepared when it entered the war in 1917. It recruited and organized electricians and utility specialists, creating an Engineer Reserve Corps in 1916. The Illuminating Engineering Society that in peacetime promoted white ways and electric advertising helped to find qualified electricians for the military. Nevertheless, the US military underestimated the demand for electricity along the western front. It found it difficult to acquire equipment in Europe (not least because the electrical systems were incompatible), making it necessary to import items from the United States. The military set up a large telephone system centered in Tours, supplying it entirely with US equipment. The Americans also wired and supplied electricity to 150 hospital buildings with a thousand beds in each, as well as larger facilities, some with five thousand beds. In short, just as the city had become a complex networked system during the nineteenth century, a modern battleground also demanded extensive electrical services.

When the war ended, crowds celebrated all night in Paris, where the hotels and other buildings "were brilliantly illuminated by gas and electricity for the first time in five years."[65] In the United States, too, "artificial light played a prominent part throughout the country in the joyful festivities." A "jeweled archway, inspired by the Panama Pacific Exposition," greeted soldiers returning to New York City. The arch hung like a curtain of jewels between two obelisks, which rose to a height of eighty feet, surmounted by sunbursts.[66] It contained roughly thirty-two thousand "gems" of the same kind used in San Francisco, "prisms in ruby, jonquil, olive, and ultramarine blue" that sparkled as "great beams of light" played over them from "several dozen searchlights" with a collective candlepower of ninety-six million.[67] Chicago created a "Victory Way" of red, white, and blue lamps, with a "golden goddess of victory" rising above them. In the center of this illumination stood "an Altar of Victory," consisting of two ninety-foot candelabra, "studded with jewels." Each candelabra had five "candles" made from red and orange lamps, and each poured out illuminated steam, creating the impression that the candles were alight (see figure 8.8). The candelabra held up "a curtain of jewels." In the middle of this "curtain," a jeweled American eagle held the flags of the Allies. Projectors with two hundred million candlepower illuminated the whole display, including, above the candelabra, a gigantic fan of searchlight beams.[68]

In July 1919, Washington held an International Festival of Peace, organized by the War Camp Community Service. During the war, the service had staged many events for soldiers, with assistance from the Red Cross and Daughters of the American Revolution. The festival began at 5:00 p.m. and used increasingly spectacular lighting effects as evening turned into night. First the public viewed a series of tableaux, where women in Greco-Roman clothing solemnly represented the "Call of Liberty," "Call of the Children," "Call of the Land," and other abstract representations. Then came a parade of floats from many nations. The French float represented the recovery of Alsace-Lorraine, the Dutch depicted Henry Hudson's ship, the British float "Britannia" staged a maypole dance, and so forth. Next, in a pageant on the steps of the Capitol, women dressed as allegorical figures showed how love triumphed over hate, and peace over war. Then a figure representing the United States descended the steps to lift up figures representing oppressed nations, escorting them to meet a woman dressed as "Liberty." As in Bragdon's Song and Light extravaganzas in Central Park, a chorus of a thousand voices accompanied these allegories. This climaxed with

8.8 Altar of Jewels Celebrating End of World War I, Chicago
Source: Hall of History, Schenectady, NY

women representing different countries dancing together as part of the new League of Nations. Throughout the pageant, lighting dramatized the central figures. Immediately after the pageant came Ryan's fireless fireworks and special effects, using the display technologies deployed at the Hudson-Fulton Centennial and Panama-Pacific Exposition.[69]

That summer, the electrical utilities were vigorously promoting a return to full public illumination, and they organized a "brighten up" campaign that spread to most cities and towns. A "glowing example was set in Washington, where the floodlighting of the Capitol" returned after being discontinued during the war.[70] The city as a whole was "bathed in light," again under Ryan's direction. Army engineers attracted thousands of onlookers, as they used "gigantic searchlights" that "sent a glow through the heavens easily visible in Maryland and Virginia within 50 miles of Washington." The public was encouraged to bring cameras, as the lighting of prominent buildings was so intense that night photography would be possible. A low-flying airplane "made the scene additionally picturesque, the machine weaving its way in and through the great shafts of light that were directed against the White House and other well-known buildings." Immediately afterward, the National Press Club sponsored an entertainment with the same theme as the San Francisco parade of 1916—namely a "pictorial history of illuminations from the time of the cave man to the present day." The theater was packed with dignitaries, including members of the Cabinet, House, and Senate.[71]

Curiously, US developments moved in the opposite direction from those in Britain. There, as Alice Barnaby explains, illuminations had been contentious and highly political in the eighteenth and early nineteenth centuries, but then attenuated into a few "ceremonial occasions of state such as births, deaths and marriages of monarchs." This "depoliticization," she argues, left brilliant lighting to commerce, which commodified experience, notably at amusement parks.[72] Britain's streets also remained largely in the gas era, compared to the United States, where during the same years lighting radically increased and, as this chapter has emphasized, became far more political.

Unlike London, Washington had never been a leader in acquiring gas or electric lighting. It had never staged a world's fair, and the city had little electrical advertising compared to Chicago or New York.

The impulse to develop spectacular lighting had come almost entirely from other cities and large corporations. Other US cities developed forms of lighting that attracted the public, relying on the aesthetics of both the Great White Way and expositions. When Americans began to celebrate military victory with illuminations, the technologies chosen emerged from corporate mass culture more than from political traditions. The result can be considered hegemonic, but it was a form of cultural dominance largely based on popular enthusiasm for the new technology of electrical spectacles. During the eighteenth century, European courts had employed experts in fireworks and expected them to innovate new forms of entertainment. The equivalent US experts were not government employees but instead worked for General Electric, Westinghouse, and other corporations, and their techniques had been perfected at the Hudson-Fulton Celebration, Panama-Pacific Exposition, Coney Island, and Times Square. Not only the technologies of display but also much of the imagery came to Washington from such venues, notably the idea that the spread of light symbolized the advance of civilization. That theme was familiar to the millions who attended the expositions in Omaha, Buffalo, Saint Louis, and San Francisco. The brightly illuminated Statue of Liberty delivered the same message, identifying the nation and democracy with its torch of freedom.

Spectacular lighting had become a US prerogative of power. Electrical effects illuminated a patriotic landscape that included the Washington Monument, White House, US Capitol, Statue of Liberty, Grant's Tomb, Niagara Falls, Brooklyn Bridge, and many other sites, including the scintillating skylines of major cities. Spectacular lighting had also become an essential part of ritual events, such as the Fourth of July, and standard methods of display had evolved that included flashing lighting arrays, powerful searchlights, electric fountains, electric American flags, and jeweled arches. Public lighting, once of little interest to the government in Washington, had become an integral part of political culture, in parades, commemorations, rallies, bond drives, election campaigns, conventions, victory celebrations, and inaugurations. The United States and Britain had evolved different lighting practices. On both sides of the Atlantic, culture shaped technology.

9

Multiple Blindings

Even as the national government adapted the spectacular lighting of world's fairs to political purposes, the first era of great expositions ended. After the Panama-Pacific Exposition closed, workers tore up its sidewalks, revealing railroad tracks beneath them that they used to cart away rubble from its demolition. In three months, the great exposition had disappeared, and there would be no more great US expositions for a generation. They would be resurrected in the 1930s, but by then the vision of a Beaux-Arts city had largely disappeared. The San Francisco fair proved to be not a harbinger of the future but rather the last major attempt to convince Americans to embrace the City Beautiful movement. The fair also marked the end of the transition from gas to electricity that had begun in the 1870s. By 1915, electrification had achieved technological momentum. The electrical utilities had grown into mature, profitable local monopolies, and electricity had become the preferred source of light. Electrical manufacturing was dominated by the duopoly of General Electric and Westinghouse. As the energy transition drew to a close, the utopian expectations that animated fairs from 1881 until 1915 faded, leaving a residual faith in progress.

During the period of transition there had been great variety in urban illumination. US cities and towns were laid out as grids that tended toward a monotonous homogeneity, and after 1880 they sought variety through lighting. For several decades, variations in illumination became part of the charm of travel. When Preece went from New York to London in 1884, he was struck by how much darker it was in his native England. When Twain visited Detroit in 1884, he enjoyed the novelty of its tower lighting. When the young newspaper reporter Dreiser went from gaslit Saint Louis to see Chicago's Columbian Exposition, he was stunned by its display of electrical lighting. When he later visited New York, Dreiser encountered the novelty of monumental advertising signs. As late as 1900, when Cincinnati sent a committee to inspect the lighting systems of ten other cities, they found wide variation in the systems, intensity of lighting, and placement and appearance of the standards. Milwaukee was saturated with the sepia coloration of gaslight, while nearby Chicago preferred the brighter but cooler tonality of electric arc lights. Cities had such different kinds and intensities of illumination that travelers were forcibly struck by the contrasts. Everywhere, however, comparatively little light came from the heavens, as smoke pollution blotted out the stars and reduced the moon to pale inconsequence.

Europe also witnessed culturally driven variations in public lighting. Beaumont has argued that in nineteenth-century Britain there were two aesthetics of light. Keats and many others embraced Romantic darkness, "the rich, gloaming gloom of the garden at night," when "the distinction between the environment and the individual or organism at its centre momentarily seems to disappear."[1] In contrast, another aesthetic embraced the powerful effects that electrification made possible and considered them sublime. These contrasting choices were uneasily reconciled in the silvery dimness of moonlight or mellowness of gaslight, which was retained in London a generation longer than in Chicago or Berlin. Spectacular lighting mesmerized many a visitor to Paris or Berlin, and Dickens complained that London was too dark. Yet Stevenson wrote an essay in praise of gaslight, which in Britain "quickly came to seem mysterious and magical." As Lynda Nead concluded, the gaslit city of London "became a place given over to imagination, dread and dream."[2] In contrast, some of the newer US cities never even installed gas, and most of those that did soon began to install arc lighting. "Moonlight" electric towers spread widely in the United States during the 1880s, but lasted barely a decade before being replaced with more conventional arc lights. These too were quickly abandoned after 1910 in favor of tungsten incandescent lights, successfully marketed as white ways all over the country (see figure 9.1).

Electric lighting was the entering wedge of an energy transition that was completed a generation earlier in the United States than in

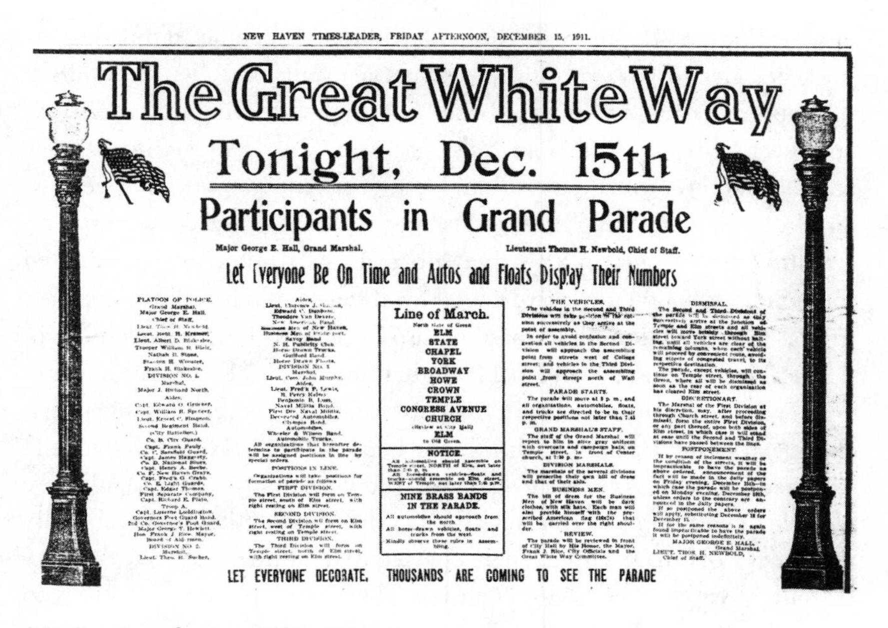

NEW HAVEN TIMES-LEADER, FRIDAY AFTERNOON, DECEMBER 15, 1911.

The Great White Way

Tonight, Dec. 15th

Participants in Grand Parade

Major George E. Hall, Grand Marshal. Lieutenant Thomas H. Newbold, Chief of Staff.

Let Everyone Be On Time and Autos and Floats Display Their Numbers

PLATOON OF POLICE.
Grand Marshal,
Major George E. Hall.
Chief of Staff,
Lieut. Thos. H. Newbold.
Lieut. Robt. H. Kramer,
Lieut. Albert D. Blakeslee,
Trooper William H. [illegible],
Nathan B. Stone,
Stanton H. Webster,
Frank H. Blakeslee.
DIVISION NO. 1.
Marshal,
Major J. Richard North.
Aides,
Capt. Edward O. Gruener,
Capt. William R. Speckert,
Lieut. Ernest C. Simpson.
Second Regiment Band.
(City Battalion.)
Co. B. City Guard,
Capt. Frank Pauly.
Co. C. Sarsfield Guard,
Capt. James Haggerty.
Co. D. National Blues,
Capt. Henry A. Beebe.
Co. F. New Haven Grays,
Capt. Fred'k G. Crabb.
Co. E. Light Guards,
Capt. Edgar Thomas.
First Separate Company,
Capt. Richard E. Plato.
Troop A.
Capt. Laverne Luddington.
Governors Foot Guard Band.
2nd Co. Governor's Foot Guard,
Major George T. Hewlett.
Hon. Frank J. Rice, Mayor,
Board of Aldermen.
DIVISION NO. 2.
Marshal,
Lieut. Theo. H. Sucher,
Aides,
Lieut. Clarence J. [illegible],
Edward C. Dunham,
Theodore Van Desen,
New American Band
Business Men of New Haven,
Business Men of Bridgeport.
Savoy Band
N. H. Publicity Club
Horse Drawn Trucks.
Guilford Band.
Horse Drawn Floats.
DIVISION No. 3
Marshal,
Lieut. Com. John Murphy.
Aides,
Lieut. Fred'k P. Lewis,
R. Percy Kelsey,
Benjamin B. Lum,
Naval Militia Band.
First Div. Naval Militia.
Decorated Automobiles.
Champia Band.
Automobiles.
Wheeler & Wilson Band.
Automobile Trucks.

All organizations that hereafter determine to participate in the parade will be assigned positions in line by special orders.

POSITIONS IN LINE.

Organizations will take positions for formation of parade as follows:

FIRST DIVISION.

The First Division will form on Temple street, south of Elm street, with right resting on Elm street.

SECOND DIVISION.

The Second Division will form on Elm street, west of Temple street, with right resting on Temple street.

THIRD DIVISION.

The Third Division will form on Temple street, north of Elm street, with right resting on Elm street.

Line of March.
North Gate of Green
ELM
STATE
CHAPEL
YORK
BROADWAY
HOWE
CROWN
TEMPLE
CONGRESS AVENUE
CHURCH
(Review at City Hall)
ELM
to Old Green.

NOTICE.
All automobiles should assemble on Temple street, NORTH of Elm, not later than 7:0 p. m.
All horse-drawn vehicles—floats and trucks—should assemble on Elm street, WEST of Temple, not later than 7:0 p.m.

NINE BRASS BANDS IN THE PARADE.
All automobiles should approach from the north.
All horse-drawn vehicles, floats and trucks from the west.
Kindly observe these rules in assembling.

THE VEHICLES.

The vehicles in the Second and Third Divisions will take position in the column successively as they arrive at the point of assembly.

In order to avoid confusion and congestion all vehicles in the Second Division will approach the assembling point from streets west of College street; and vehicles in the Third Division will approach the assembling point from streets north of Wall street.

PARADE STARTS.

The parade will move at 8 p. m., and all organizations, automobiles, floats, and trucks are directed to be in their respective positions not later than 7:45 p. m.

GRAND MARSHAL'S STAFF.

The staff of the Grand Marshal will report to him in olive gray uniform with overcoats and campaign hats, on Temple street, in front of Center church, at 7:30 p. m.

DIVISION MARSHALS.

The marshals of the several divisions will prescribe their own bill of dress and that of their aids.

BUSINESS MEN.

The bill of dress for the Business Men of New Haven will be dark clothes, with silk hats. Each man will also provide himself with the prescribed American flag (16x24) that will be carried over the right shoulder.

REVIEW.

The parade will be reviewed in front of City Hall by His Honor, the Mayor, Frank J. Rice, City Officials and the Great White Way Committee.

DISMISSAL.

The Second and Third Divisions of the parade will be dismissed as they successively arrive at the junction of Temple and Elm streets and all vehicles will move briskly through Elm street toward York street without halting, until all vehicles are clear of the remaining columns, when each vehicle will proceed by convenient route, avoiding streets of congested travel, to its respective destination.

The parade, except vehicles, will continue on Temple street, through the Green, where all will be dismissed as soon as the rear of each organization has cleared Elm street.

DISCRETIONARY.

The Marshal of the First Division at his discretion, may, after proceeding through Church street, and before dismissal, form the entire First Division, or any part thereof, upon both sides of Elm street, in which case it will stand at ease until the Second and Third Divisions have passed between the lines.

POSTPONEMENT.

If by reason of inclement weather or the condition of the streets, it will be impracticable to have the parade as above ordered, announcement of the fact will be made in the daily papers on Friday evening, December 15th—in which case the parade will be postponed on Monday evening, December 18th, unless orders to the contrary are announced in the daily papers.

If so postponed the above orders will apply, substituting December 18 for December 15.

If for the same reasons it is again found impracticable to have the parade it will be postponed indefinitely.

MAJOR GEORGE E. HALL,
Grand Marshal.
LIEUT. THOS. H. NEWBOLD,
Chief of Staff.

LET EVERYONE DECORATE. THOUSANDS ARE COMING TO SEE THE PARADE

9.1 Opening of a Great White Way
Source: Hall of History, Schenectady, NY

most of Europe. As the US electrical system penetrated into all sectors of society, utilities were better able to "balance the load." Because their generating systems were in more constant use, they were more efficient and offered lower electricity prices, which in turn fostered further demand. The technological momentum of the US electrical system was embodied not only in massive lighting arrays but also in hundreds of subsidiary industries that manufactured appliances, university curricula that trained thousands of electrical engineers, workshops that handcrafted large electrical signs, newly electrified schools and factories, and the associations and interest groups that knit together the industry, such as the American Society of Electrical Engineers, National Electric Light Association, and Society of Illuminating Engineers.

While professionals consolidated the electrical supply system, the public also had its say. People decided whether to light their streets and businesses with gas or electricity, and what systems they preferred. (In the 1920s, they would decide how much and what kind of light they wanted in their kitchen, bedroom, or basement, and whether they wanted electric lights on their Christmas tree.) Millions of people attended lighting spectacles, first at expositions and special events, and then at hundreds of amusement parks, which splashed the night with ingenious displays and special effects. Most Americans considered Times Square not an eyesore but rather a tourist attraction. Despite attempts to curtail spectacular electrical advertising, that industry survived the attacks of the City Beautiful movement and became "normal." Most downtowns adopted the white way, and cities were zoned for different levels of light, depending on their function. Illuminations also became an essential element in political campaigns, presidential inaugurations, and public monuments.

These decisions made US urban landscapes even less like their European counterparts than they had been in 1875. Not only was the US metropolis laid out as a grid. It was more intensely lighted, had more electric advertising, and erected taller buildings (themselves heavily dependent on electricity in their operation), and these skyscrapers were extremely visible due to floodlighting. There were partial exceptions to these generalizations, notably Boston, where the grid pattern was less fully realized than elsewhere and the skyscraper was long resisted.[3] Yet Boston had the most intense street lighting in the country.

The subjective experience of these public spaces varied by gender and class. When dark, the night city had largely been a male space, unsafe for women, especially if unaccompanied. Wealthy individuals could travel in coaches with accompanying servants, which gave them security and greater freedom of movement than others. As the city became better lighted, it was safer not only because it was brighter but also because more people ventured out. Illuminated cityscapes opened up spaces for unchaperoned women, so that they might enjoy more freedom of movement, but this freedom was greatest in commercial zones. As William Leach found in his work on department stores, women gained greater urban mobility.[4] Women could not roam safely everywhere, however; they were most secure in central shopping districts. There was a safety in numbers from armed robbery, rape, and assault, although a crowd gave pickpockets opportunities and afforded strangers occasions for unwanted familiarity. Yet overall, the lighted city was a safer place that encouraged social contact. These effects reinforced one another. When there were more vaudeville shows, movie houses, and theaters, their audiences patronized more restaurants, taverns, and cafés. The more people were out, the more businesses remained open, encouraging further throngs to take to the streets. But if social classes mixed in commercial zones, lighting was more lavish in wealthy neighborhoods than in impoverished ones, where gas lighting often lingered.

The flannêur was hardly visible in this electrified landscape, where the central experience was not walking alone but instead moving with a crowd that sought dynamic experiences of space. People rode in a packed elevator to the top of a skyscraper and gazed at the urban panorama from the observation deck. They took electrified streetcars, screamed together on roller coasters, jammed into dance halls, applauded or booed at the theater, or bicycled in clubs. Stores installed large plate glass windows and well-lighted displays to attract crowds on their way to the theater, vaudeville show, or the new movie houses. This urban crowd made locations such as New York's Broadway or Chicago's Loop worthwhile sites for national advertisers, who erected enormous signs that became attractions in themselves. In a self-reinforcing process, increased lighting drew larger crowds into public space, encouraging further increases in lighting. This process reached its apogee in the second decade of the twentieth century, when the interiors of most homes and apartments were still unelectrified and darker than the city street. Accordingly, most entertainment took place in public space as opposed to the privacy of the home. This slowly began to change once millions of homes were electrified in the 1920s.[5] The lure of the street slowly weakened as the radio and phonograph (and television one generation later) provided alternatives.

The attractions of the electrified city also started to face competition from automobiles. Car ownership spread rapidly after circa 1910, as improved production methods lowered the price tag.[6] By the end of the 1920s there was on average an automobile for every family in the United States, and people drove these vehicles to new areas of consumption that sprang up on the edge of town. The noise and pollution of automobiles made central cities less attractive, and drivers wanted lighting designed to increase the visibility of the roadways. These sheets of light often lacked subtlety and did not always create an attractive walking environment. By the end of the 1920s fewer people took mass transit into the center of town. The trolley lines and their hundreds of amusement parks slowly went bankrupt. The city was more brightly lighted than ever, but the crowd's attention had begun to shift elsewhere.

Gearing up to meet the needs of motorists, in 1914 the National Electric Light Association joined with the Association of Edison Illuminating Companies to establish parameters for street lighting. They agreed that outdoor urban lighting had to illuminate the street and help people to keep their orientation, but the level of light sufficient for walking was no longer adequate. Automobile traffic dictated lighting that allowed safe movement at a much higher speed than a horse-drawn carriage. "No longer adequate was the pooling of light at intervals along streets. A strictly even illumination was now needed, in which vehicles and pedestrians could be fully lit. … No longer could city lighting systems function only on moonless nights or only until midnight."[7]

General Electric recommended dividing cities into six lighting zones. The lumens per square foot in the central business district were six to eight times stronger than in residential sections.[8] This hierarchical system expressed an altogether-different conception of the city than the egalitarian moonlight towers of the 1880s or aestheticism of the City Beautiful movement. It prioritized businesses and motorists, and valued pedestrians primarily as shoppers. Table 9.1 only includes public lighting. If advertising signs and shop windows were also included, the difference between a white way and the rest of the city would be even greater. General Electric and the National Electric Light Association saw each town as a nodal point in a larger "web that will help knit the nation together by turning into ribbons of light the highways that lead out through the open country to distant cities."[9]

This hierarchy had a higher purpose. As John A. Jakle has noted, lighting the street was not only useful for navigation or selling goods in store windows but also "could advance city pride, demonstrate competent city administration, and foster other civic improvements."[10] Intensive lighting defined the desirable city. As a 1916 General Electric advertisement emphasized, lighting made "your streets brighter, your town prosperous, your homes safe, [and] your living conditions better." It made "your city more attractive, healthier, busier, [and] cleaner." It had become "indispensable to a wholesome town, a 'live' town, a happy town, a good town."[11]

Table 9.1
Classification of Urban Street Lighting, 1915

Class of street	Lamps to be used (lumens)	Mounting height, feet	Feet of street per standard	Lumens per foot of street
Intensive white way	15,000 and 25,000 luminous arcs	18–25	50–75	300–800
White way	10,000 and 15,000 luminous arcs	14–16	40–65	200–400
White way, secondary	6,000–15,000 luminous arcs	14–16	40–65	125–250
Main thoroughfare	4,000–10,000 incandescents	14–16	50–75	80–150
Secondary thoroughfare	2,500–6,000 incandescents	12–14	50–75	50–80
Residential	2,500–4,000 incandescents	11–14	60–100	25–50

The hierarchies that a zoned system of lighting established did not benefit all citizens equally. Just as the "barbarous races" were exhibited in primitive villages on the outskirts of the great expositions, city lighting schemes frequently marginalized the poor, Chinese and Japanese immigrants, and blacks. Research is scant, but it appears that electricity only reached the urban poor and minorities when building codes required it. Municipal tower lighting had been more egalitarian, providing the same level of light for all citizens. The newer, magnificent illuminations were almost entirely conceived and built by white men during the period when Americans developed an intricate racial hierarchy that was expressed by segregation for blacks, forced assimilation for Native American children, and the curtailment of Chinese and Japanese immigration. This hierarchy also was registered in the layout of world fairs and differential illumination of cities.[12]

Furthermore, intensively lighted business districts had a competitive advantage over the smaller stores in ethnic neighborhoods. The new electric streetcars and subways concentrated shopping in the center of town. There, department stores offered a larger selection of goods and could sell at a lower price because of the rapid turnover that came with the swelling ranks of customers. As commerce concentrated in the center and traffic intensified, the streets lost some civic functions and were reconceived as arteries of transportation between functionally defined urban zones. Jacobs later attacked this conception of the city, which decreed that one should not live where one worked, shopped, or found entertainment. She preferred a more European pattern, with multiple uses of urban space, so that people lived near their work, shops were not segregated from residences, and one could obtain most necessities by walking or bicycling.[13] The use of space in the preelectric era had been multifunctional, and different social classes had then lived in closer proximity. US cities used electrification (as well as other technologies) both to create a scintillating downtown where all social classes met and to differentiate residential space, which increased social distance.

During the energy transition from gas to electricity, lighting technologies were in rapid evolution including DC arc lighting, tower lighting, incandescent carbon filament lighting, AC arc lights, Welsbach gas mantles, the improved GEM carbon lamp, and the tungsten filament lamp (see pages 249 to 251). By 1910, these were joined by brillant floodlights and searchlights, and the innovation of fireless fireworks that used powerful projectors to beam images and patterns into fog or smoke generated for special occasions. Yet technologies did not drive change—people did. Many actors shaped the aesthetics of this new landscape, including government, local elites, businesses, utilities, and the large corporations that sold the systems. Most obviously, Westinghouse and General Electric pushed the development, display, and sale of new forms of lighting. They employed inventors and lighting engineers who pioneered new forms of public illumination. They worked in partnership with local utilities.[14] The utilities saw each exposition or special event as an opportunity to build demand for electricity, and temporary special effects were a means to that end. Local businesspeople often led agitation for improved lighting, and at times paid for and installed street lighting systems in order to draw as well as hold customers. In such cases, once Main Street had more intense illumination, businesspeople usually turned the system over to the city, which often had only marginal influence on the placement or purpose of the lights. Otherwise, city engineers usually made these choices. Utilities promoted street and commercial lighting, but before 1900 they only sporadically considered the overall aesthetics of the electrical landscape. The public also played a role, both as voters and consumers. Did people want gas or arc lighting, tower or pole lighting, public or private ownership, or a uniform or individualistic aesthetic? In contrast to the lively incoherence of downtowns, World's fairs created harmonious aesthetic patterns. This inspired local civic groups and the City Beautiful movement, which at times managed to restrict the size of advertising signs or install ornamental lighting poles.

By 1905, the genteel elite and some of the middle class wanted an electrical landscape where unsightly details had been excised. They wanted illuminations that exalted neoclassical buildings and the horizontal city of Beaux-Arts architects, in opposition to the emerging vertical, skyscraper city. Such ideas culminated in the Hudson-Fulton Celebration of 1909 and Panama-Pacific Exposition of 1915. Expositions imposed a progressive order on fairgrounds, where electric light, heat, power, and transportation helped depict humankind's rapid evolution from savagery to civilization toward a blueprint of the future. Such progress seemed especially palpable at expositions held where a century before there had scarcely been a town, as was the case with Chicago, Omaha, Saint Louis, Seattle, San Diego, and San Francisco.[15]

Americans had decided by 1915 that the night city should not resemble the city by day. The tower lighting of the 1880s strove to present the urban landscape as a softer yet complete version of its daylight self. Cities used arc lights to create artificial moonlight that would illuminate not individual buildings but rather the city as a whole. The City Beautiful movement took up related ideas, as did the planners of major expositions and civic events. Yet Americans tended to reject comprehensive lighting schemes. Instead, much of the metropolis was literally relegated to the shadows. Areas of poverty disappeared at night, as did any business that chose not to pay for extra lighting to advertise itself. The resulting landscape expressed the competitive values of private enterprise.

Commercial lighting articulated without words a good deal of the US political-economic system, with its tensions between rural and urban, art and commerce, wealth and poverty, and public and private space. The cityscape that resulted was neither as uplifting as the City Beautiful movement wanted nor as boisterous and visually hyperactive as Coney Island or the midway of a world's fair. This electrified city lacked the visual unity of a Beaux-Arts exposition, and it threatened to devolve into many disconnected social worlds, each with its own identity and visual aesthetic. The rejected alternative of tower lighting had pointed toward a city that was less commercial, more coherent, and legible—a city that minimized the alternation between its day and night appearance. The city had instead been defamiliarized. Streets and buildings changed their appearance at night, when the urban landscape pulsated with new energies. The downtowns had become heterotopian spaces that transformed themselves every night through lighting. Americans had developed an almost dematerialized and idealized city of possibility—one that outlined, highlighted, or spotlighted select locations. This seemingly enchanted city also defined its opposite. If the capitalist nocturne blotted out pollution, poverty, blight, and cultural differences, these repressions exoticized "the other half" faintly visible in the shadows. At the center of this system of illumination stood paradigmatic structures that became symbols of US society. As Luckiesh, a leading lighting engineer, put it, "Just as the Statue of Liberty stands alone in the New York Harbor so does the Woolworth Building reign supreme on lower Manhattan. Liberty proclaims independence from the bondage of man and the Woolworth Tower stands majestically in defiance of the elements as a symbol of man's growing independence of nature."[16] Both were bathed in powerful white light and stood against the night sky. Other cities highlighted skyscrapers and landmarks that played a similar role.

Around these symbols, the US city changed in accord with whatever most attracted and stimulated the public. In the new zones of illuminated night there were strong contrasts, strange shadows, and little sense of depth. Rows of streetlights provided perspective lines, but these were disrupted by lighted billboards and flashing advertising signs that were not designed to a common scale. The confused relations between front and back turned the central space of the city into a pulsating visual collage. The sheer power of the lighting arrays, combined with air pollution, blotted out most of the stars and dimmed the moon. Instead, illuminated landmarks provided visual orientation. The uncertain sense of scale was both exhilarating and confusing. The US metropolis had become a vibrant, indeterminate text that changed with each new building and every new flashing sign, to the frustration of the urban planners who sought an orderly, legible city, but to the delight of visitors to the Great White Way. To some, this night city was sublime. It seemed to synthesize many discordant elements into a dynamic whole, as it celebrated the towers, monuments, bridges, railway stations, and boulevards. This New World landscape gave vent to market forces, as skyscrapers and enormous signs vied for attention and advertising pervaded public space. The landscape became the visual correlative of individualistic commerce, and its lighting embedded dynamism and competitive values in the city's fabric. After 1910, illumination would begin to serve automobiles more than pedestrians, particularly as one moved away from the center. As speeds increased, the city was edited to be legible to fast-moving traffic.

This electrified landscape was made possible by coal mines, oil wells, railroads, hydroelectric dams, and vast factories producing equipment. It required an infrastructure of high tension lines, power plants, cables under the streets, and hidden wires that grew more intricate with the widespread adoption of each new convenience. After 1920, the electrical system achieved so much technological momentum that it began to seem an inexorable historical force, inevitable transformation, even a kind of fate. The glittering sea of lights in large US cities became a visual metaphor for "the electrical age." The cities of light seemed to epitomize the triumph of technological civilization. The electrified city represented the earth's subjugation.

In retrospect, the early twentieth century was a moment of hubris, for its extensive use of gas and electricity accelerated global warming. Permanent illumination was not cost free. It would require a century to discover and understand the environmental consequences, however. A satellite circling the earth since 1000 AD would have recorded the sudden burst of illumination after 1800, its intensification as gas lighting gained technological momentum, and a second brightening when arc and incandescent lights started to spread after 1880. Such a satellite might also have noted the gradual warming of the earth as more carbon dioxide was released. It might have recorded the rise in air pollution, acidification of lakes, deforestation, and rising seas caused in good part by coal-fired electrification. But there was no such satellite and no overview.

Walter Benjamin concluded that an "overabundance of light produces multiple blindings."[17] This was literally the case for anyone looking at a 2000-candlepower arc light, but it was also an accurate metaphor in many other senses. Both gas and electric light distorted the appearance of objects, because they by no means replicated the colors of sunlight. As used in illuminations for special events, beginning in the Renaissance, powerful lighting was a hegemonic form of social power. As gas and later electric light were dispersed in society, they were seldom used to light all areas equally. Instead, light often remained hegemonic, and was used to promote some locations and cast others into the shadows. By the end of the energy transition, Americans had become almost blind to whatever was not illuminated as well as to the structures of power that had zoned the city and directed their gaze to flashing signs, corporate headquarters, monuments, public buildings, and patriotic symbols. As these artificial forms of lighting were naturalized, Americans also became blind to alternatives. The "overabundance of light" had become both an emblem of modernity and an apparent release from nature's rhythms and limits. In 1915, whether strolling on the new white way in a city center, watching a parade, visiting one of the great expositions, gazing from the observation platform of a skyscraper, caught up in the synesthesia of an amusement park, or driving through the streets in one of the newfangled motor cars, the public was mesmerized by American illuminations.

Notes

Introduction

1. Nye, *Electrifying America*, 238–286.

2. Schivelbusch, *Disenchanted Night*, 75–76.

3. Nye, *Electrifying America*, 199–200.

4. Nye, *America's Assembly Line*, 18, 27; Scranton, *Endless Novelty*, 129.

5. Nye, *Technology Matters*, 49–66.

6. Jones, *Routes of Power*, 2–21, 162–167.

7. Schivelbusch, *Disenchanted Night*, 3, 73, 74.

8. "Table 13: Population of the 100 Largest Urban Places: 1900," US Bureau of the Census, June 15, 1998, accessed May 3, 2017, https://www.census.gov/population/www/documentation/twps0027/tab13.txt. Some major cities of 2015 were much smaller in 1900, and therefore receive little attention in this book, notably Los Angeles (102,000), Atlanta (90,000), Houston (45,000), Dallas (43,000), and Phoenix, which had less than 38,000 inhabitants.

9. Bouman, "The Good Lamp Is the Best Police," 63–78; Leach, *Land of Desire*.

10. Wells, *The Future in America*, 48.

11. Stradling and Thorsheim, "The Smoke of Great Cities," 6–31.

12. See chapter 7.

13. Pound, "Patria Mia," 107.

14. Michel Foucault, "Of Other Spaces: Utopias and Heterotopias," *Architecture/Mouvement/Continuité*, March 1967 lecture, trans. Jay Miskowiec, accessed May 3, 2017, http://web.mit.edu/allanmc/www/foucault1.pdf.

15. Ibid.

16. "The Electrical Utility Exhibits at the New York World's Fair," 81–87.

17. Lears, *No Place of Grace*, xvi.

Chapter 1: Illuminations

1. Biringucci, *La Pyrotechnie*, 168. Schivelbusch dates illuminations to the baroque period. Schivelbusch, *Disenchanted Night*, 137–138.

2. Ruggieri, *Elemens de pyrotechnie*, xvi (my translation).

3. Werrett, *Fireworks*, 133.

4. Goethe, *Italian Journey*, 344.

5. Werrett, *Fireworks*, 107–108.

6. Bressani, "Paris," 28.

7. Lynn, "Sparks for Sale," 77.

8. Koslofsky, *Evening's Empire*, 131–132.

9. Habermas, "The Public Sphere," 49–53.

10. O'Dea, *Social History of Lighting*, 95.

11. Burrows and Wallace, *Gotham*, 111.

12. Ekirch, *At Day's Close*.

13. Ibid., 177.

14. Ibid., 227–230.

15. Koslofsky, *Evening's Empire*, 131–132.

16. Brockett, *History of the Theater*, 201, 297.

17. Koslofsky, *Evening's Empire*, 95.

18. Lynn, "Sparks for Sale," 74–75.

19. Koslofsky, *Evening's Empire*, 101, 157.

20. *Flying Post* or *Post Master* (London), December 23, 1699; January 14, 1701; *Post Boy* (1695) (London), December 18–20, 1711, issue 2591.

21. *Daily Courant* (London), December 28, 1702, issue 217; November 19, 1703, issue 497.

22. Koslofsky, *Evening's Empire*, 96.

23. Goulet, *Fêtes à l'occasion du mariage des S. M. Napoléon, empereur des Français, roi d'Italie*.

24. Neumann, *Architecture of the Night*, 99.

25. Ibid., 9.

26. Twain, *The Innocents Abroad*, 219.

27. Twain, *Historical Romances*, 59.

28. Kaempffert, *Ornamental Street-Lighting*, 32.

29. Lynn, "Sparks for Sale," 77–78, 85–86.

30. Ibid., 86–87.

31. *Morning Post* (London), November 8, 1805.

32. *London Examiner*, Sunday, July 11, 1813.

33. Tomory, *Progressive Enlightenment*, 212–225.

34. Mortimer, *Pyrotechny*, xiv. Copy, Dibner Library, Smithsonian Institution, Washington, DC.

35. Monroe, *Narrative of a Tour of Observation Made during the Summer of 1817*, 121; "President's Tour," *Niles Weekly Register*, August 9, 1817, 371.

36. Text included on the print: "View of the Magnificent and Extraordinary Fireworks exhibited on the N.Y. City Hall on the Evening of the celebration of the Grand Canal, November 4, 1825." New York Public Library, accessed May 4, 2017, https://digitalcollections.nypl.org/items/510d47da-28c6-a3d9-e040-e00a18064a99.

37. On celebrating peace, see "Wakefield Illumination," *Leeds Mercury*, June 11, 1814; "Illumination at Soho Manufacturing at Birmingham, June 9, *Morning Chronicle* (London) June 23, 1814. On celebrating a royal visit, see *Caledonian Mercury* (Edinburgh), August 22, 1822.

38. "The General Illumination," *Brighton Patriot and South of England Free Press*, Tuesday, October 10, 1837.

39. Ibid.

40. "German Visitors on Illumination Night," *Leeds Mercury*, September 20, 1855.

41. Cited in Werrett, *Fireworks*, 203.

42. Dodderer-Winkler, *Magnificent Entertainment*, 119–131.

43. Barnaby, *Light Touches*, 114.

44. "Illumination and Riots at Edinburgh," *Glasgow Herald*, November 20, 1820.

45. Ibid.

46. *Bury and Norwich Post*, or *Suffolk, Essex, Cambridge, Ely, and Norfolk Telegraph*, August 1, 1821.

47. "The Illumination," *Freeman's Journal and Daily Commercial Advertiser*, May 21, 1831.

48. "Illumination Riot," *Caledonian Mercury*, March 31, 1831. See also "Illumination Riots," *Caledonian Mercury*, April 7, 1831.

49. *Morning Chronicle* (London), April 28, 1831.

50. Ibid. See also "How to Celebrate a Reform Act," *Examiner*, June 24, 1832.

51. O'Dea, *Social History of Lighting*, 178.

52. "A General Illumination—The Starving Poor," *Freeman's Journal and Daily Commercial Advertiser* (Dublin), August 1, 1849.

53. "The Illumination Riots in Ireland," *Hull Packet and East Riding Times*, March 20, 1863.

54. McNamara, *Day of Jubilee*, 16.

55. Gottfried Semper, cited in Neumann, *Architecture of the Night*, 9.

56. Bressani, "Paris," 28.

57. Ibid., 30.

58. McNamara, *Day of Jubilee*, 89.

59. Ibid., 16.

60. Seelye, "Rational Exultation," 243.

61. Nye, *American Technological Sublime*, 37–38, 47–51, 80–83.

62. "Rejoicing over the Cable," *New York Times*, August 18, 1858.

63. "Fireworks and Illuminations," *New York Times*, March 7, 1865.

64. Adams, *An Autobiography*, 69.

65. "Lincoln Parade Transparency, 1860," Natural Museum of American History, accessed May 4, 2017, http://americanhistory.si.edu/collections/search/object/nmah_513759.

66. "The Big Blaze," *Boston Daily Globe*, October 27, 1876, 1; "Last Evening's Parades," *Boston Daily Globe*, November 2, 1876.

67. "The Capitol by Gas Light" and "The White House by Moonlight," in Whitman, *Complete Poetry and Prose, Specimen Days*, 718, 757.

68. "Illuminations: Their History, the Extent of the Business, and the Improvements Made," *Boston Daily Globe*, June 23, 1878, 2.

69. "The Scene Last Night, the Illuminations," *Boston Daily Globe*, June 18, 1875, 8.

70. "Yesterday, the Great Celebration of the Century," *Boston Daily Globe*, June 18, 1875, 1.

71. "Scenes at Union Square" and "The Illuminations," *New York Times*, July 5, 1876.

72. "The National Birthday," *New York Times*, June 25, 1876, 6.

73. "Monday Night's Splendor," *New York Times*, July 5, 1876.

74. Twain, *A Tramp Abroad*, 279.

75. "History of Niagara Falls Illumination," Niagara Parks, accessed May 4, 2017, http://www.niagaraparks.com/niagara-falls-attractions/history-of-falls-illumination.html.

76. Ibid.

77. "A Grand Sight at Niagara," *New York Times*, August 14, 1881, 1.

78. Adams, *Our American Cousins*, chapter 21.

79. "History of Niagara Falls Illumination."

80. *New York Tribute*, September 5, 1907; *New York Evening Post*, September 5, 1907; *New York World*, September 9, 1907.

81. "Beginning and Development of Illuminating Engineering, Information Obtained from W. D'Arcy Ryan, Oct 27 and 28, 1925," Hammond Papers, GE Historical File, L1038–1046.

82. Nye, *American Technological Sublime*, 170–172.

83. "Festivities of Germany," *New York Times*, June 20, 1895, 1; "Royal Wedding Day Arrives," *Washington Post*, May 31, 1906, 1.

84. "Jubilee Night in London," *New York Times*, June 23, 1897, 2.

85. "All London Gay: Elaborate Preparations Made for Welcoming Troops," *Boston Daily Globe*, October 28, 1900, 22; "Coronation Illuminations," *New York Times*, June 19, 1902, 9.

86. Barnaby, *Light Touches*, 116–117.

CHAPTER 2: ENERGY TRANSITIONS

1. "How the City Is Lighted, and What It Costs," *New York Times*, February 22, 1854, 4.

2. Passer, *The Electrical Manufacturers*, 12.

3. Nye, *Electrifying America*, 29–84.

4. Tomory, *Progressive Enlightenment*; Bright, *The Electric Lamp Industry*; Passer, *The Electrical Manufacturers*, 105–129.

5. Passer, *The Electrical Manufacturers*, 195–204.

6. Baldwin, *In the Watches of the Night*, 202–203.

7. Melosi, "Energy Transitions in the Nineteenth-Century Economy," 55–67.

8. Nye, *Consuming Power*, 79–83.

9. Jones, *Routes of Power*, 6–10, 94.

10. Nye, *Electrifying America*, 26–28, 170–174, 382–384.

11. Nye, *Consuming Power*, 79–83.

12. Hughes, *Networks of Power*, 77–78.

13. Ibid., 14–17.

14. Beaumont, *Night Walking*, 339.

15. Tomory, *Progressive Enlightenment*, 239.

16. Ibid., 237.

17. Hand-colored etching by Thomas Rowlandson, "A Peep at the Gas Lights in Pall Mall."

18. Tomory, "The Environmental History of the Early British Gas Industry, 1812–1830," 29–54. See also Simpson, *Gas-Works.*

19. Beaumont, *Night Walking*, 341.

20. Schivelbusch, *Disenchanted Night*, 186.

21. Cited in Luckiesh, *Artificial Lighting*, 158.

22. Cited in Binder, "Gas Light," 363.

23. Merrick, *Report of the Committee to Whom Was Referred Sundry Memorials against Lighting the City with Gas.*

24. Cited in Schlor, *Nights in the Big City*, 244.

25. Cited in ibid., 248.

26. Sharpe, *New York Nocturne*, 42–43.

27. Otter, *The Victorian Eye*, 7.

28. Schlor, *Nights in the Big City*, 245.

29. Ibid., 235.

30. Porter, *London*, 126.

31. Dickens, *The Uncommercial Traveler*, 202.

32. Bressani, "Paris," 29.

33. Dickens, *The Uncommercial Traveler*, 202.

34. Stevenson, "A Plea for Gas Lamps," 251–252.

35. Ibid., 254–255.

36. Otter, "Cleansing and Clarifying," 56.

37. Penzel, *Theater Lighting before Electricity*, 127–152; Binder, "Gas Light," 363–365.

38. Baldwin, *In the Watches of the Night*, 16.

39. Ibid., 17–18.

40. Nye, *Electrifying America*, 261–276.

41. "A New Electric Light," *Galaxy: A Magazine of Entertaining Reading* 30, no. 3 (September 1875): 415.

42. Coopersmith, *The Electrification of Russia*, 28–31. Inwood, *City of Cities*, 281.

43. Wallace, *The Progress of the Country*, 276.

44. "The Electric Light at Paris," *Maine Farmer*, November 9, 1878, 4.

45. Clipping, *New York Times*, December 21, 1880, Hammer Papers, box 25, folder 7, Smithsonian Institution Archives, Washington, DC.

46. *American Machinist*, January 1, 1881, 8; *Scientific American*, April 2, 1881.

47. "Scientific and Useful," *New York Evangelist*, March 13, 1879, 11.

48. Schivelbusch, *Disenchanted Night*, 61.

49. Ibid. Schivelbusch relies on an inaccurate article in the *Lancet* 1 (1895): 52.

50. William Hammer, "Notes on Building, Starting, and Early Operating of the First Central Station in the World, Holborn Viaduct, London, England." Private memorandum book, Hammer Papers, box 20, folder 1, Smithsonian Institution Archives, Washington, DC.

51. Haywood, "Report to the Streets Committee, Results of the Electric Lighting of Public Ways within the City of London in 1881–1882," 22–23.

52. Ibid., 23.

53. Hammer Papers, box 25, folder 3, Smithsonian Institution Archives, Washington, DC.

54. Parsons, *The Early Days of the Power Station Industry*, 103.

55. Urbanitzky, *Electricity in the Service of Man*, 538.

56. Israel, *Edison: A Life of Invention*, 197–198.

57. Hammer Papers, box 25, folder 3, Smithsonian Institution Archives, Washington, DC.

58. "Cost of Gas and Electricity in New York," *Scientific American* 52, no. 21 (May 23, 1885): 327.

59. Bowers, *A History of Light and Power*, 112.

60. Dewey, "Street Lamps of Paris," 387.

61. "Says Boston Is Not a Sufferer, Sullivan Compares Its Lighting with Europe," *Boston Daily Globe*, October 8, 1911, 4.

62. Passer, *The Electrical Manufacturers*, 197.

63. "Electric Lights in Hartford," *Boston Daily Globe*, June 29, 1884, 1.

64. Passer, *The Electrical Manufacturers*, 197–198.

65. "Notes on Municipal Government," *Annals of the American Academy of Political and Social Science*, 200, 204.

66. Jones, *Engineering Encyclopedia*, 56–57.

67. Letter from Elihu Thomson to J. W. Hammond, River Works, West Lynn, September 7, 1932, General Electric Historical File, Publicity Department, Schenectady, New York.

68. Passer, *The Electrical Manufacturers*, 70, 67.

69. Hammond, *Men and Volts*, 303–307.

70. Luckiesh, *Artificial Light*, 160–161.

71. New York Edison Company, *Forty Years of Edison Service*, 174–175.

72. Cleveland, *Concise Encyclopedia of Energy History*, 157.

73. Passer, *The Electrical Manufacturers*, 109.

74. King, *King's Handbook*, 185–186.

75. Friedel and Israel, *Edison's Electric Light*, 146–147.

76. Whipple, *Municipal Lighting*, 50–51, 70.

77. Baldwin, *In the Watches of the Night*, 159.

78. Palmer, "Municipal Lighting Rates," 38.

79. Nye, *Electrifying America*, 265–277.

80. Byatt, *The British Electrical Industry, 1875–1914*, 25–28, 99.

81. Fri, "The Alternative Energy Future: The Scope of Transition," 6.

82. Nye, *Electrifying America*, 4–5, 26–27, 235–237, 383–384.

83. Detroit sent out a similar delegation in 1893. See "Detroit Officials Here," *New York Times*, February 26, 1893, 2.

84. Kenny, *Illustrated Cincinnati*, 45.

85. "Enclosed Arc Lamps," General Electric, 1898, Warshaw Collection, electricity series 1, box 7, Smithsonian Institution Archives, Washington, DC.

86. Committee of Board of Legislation of Cincinnati, "Report on Street Lighting in Various Large Cities in the United States," 423.

87. Ibid., 425.

88. Ibid., 426.

89. Natural gas is cleaner than coal gas, and it was known in ancient China. In the nineteenth century, it was rarely exploited. In 1821, natural gas was used in several buildings in Fredonia, New York, but it was hard to transport it to consumers. Early oil drillers discharged gas into the air or burned it off. Only in the 1870s did natural gas begin to be used extensively in towns close to gas wells, such as Muncie, Indiana. Long before pipelines carried gas from Texas and Oklahoma to eastern cities, electrification had taken over the lighting market. Nye, *Consuming Power*, 121.

90. Committee of Board of Legislation of Cincinnati, "Report on Street Lighting in Various Large Cities in the United States," 427–428.

91. Ibid., 428.

92. Ibid., 429.

93. Ibid., 450.

94. Ibid., 434–435.

95. Ibid., 436.

96. Ibid., 451.

97. As late as 1905, one family dwelling in twenty had electricity—a figure that only changed rapidly after 1918. In Britain, only 6 percent of all houses had electricity at the end of World War I. Nye, *Electrifying America*, 238–286; Hannah, *Electricity before Nationalisation*, 188.

CHAPTER 3: THE UNITED STATES AND EUROPE

1. Tarr and Dupuy, *Technology and the Rise of the Networked City in Europe and America.*

2. Schlor, *Nights in the Big City*, 86.

3. T. P. O'Connor, "The Lights o' London," *Washington Post*, October 3, 1882, 2.

4. Cited in Hughes, "British Electrical Industry Lag: 1882–1888," 38.

5. Preece, "Public Lighting in America," 66, 69.

6. Hatton, *Henry Irving's Impressions of America*, 265–266.

7. Urbanitzky, *Electricity in the Service of Man*, 562.

8. Otter, *The Victorian Eye*, 176.

9. Bowers, *A History of Light and Power*, 112.

10. "The Lighting of New York City," General Electric booklet, January 1904, Warshaw Collection, box 6, folder 7, Smithsonian Institution Archives, Washington, DC; Lacombe, "Street Lighting Systems and Fixtures in New York City," 517.

11. "Electric Lighting in Boston," *Electrical World and Engineer*, September 19, 1903, reprinted in *General Electric Company Review*, November 1903, 12.

12. "London Street Lighting," *Engineer* 80, August 2, 1895, 112.

13. Byatt, *The British Electrical Industry*, 1875–1914, 26.

14. A large correspondence took place between Jehl and Hammer. Hammer Papers, box 25, folder 5, Smithsonian Institution Archives, Washington, DC.

15. Passer, *The Electrical Manufacturers*, 94.

16. Carlson, *Tesla: Inventor of the Electrical Age*, 108–118.

17. "Historical Sketch of the Foreign Business of the General Electric Company," *General Electric Digest* 2, no. 4 (1922): 5–8; no. 5, 15–19; no. 6, 4–9.

18. Office address list, *General Electric Digest* 9, no. 6 (December 1929).

19. General Electric, "Facts about General Electric Company of Interest to Stockholders," July 25, 1932, copy in General Electric Corporation, Schenectady Library, Schenectady, NY.

20. Ibid., no. 5, 18–19.

21. As late as 1940, General Electric owned 21 percent of Osram, 34 percent of the French Compagnie des Lampes, 12 percent of Phillips, and 40 percent of Germany's AEG. See Bright, *The Electric Lamp Industry*, 309.

22. Hausman, Wilkins, and Neufeld, "Global Electrification," 175–190.

23. Koester, "Electric Lighting, Police, and Fire Alarms," 162.

24. Ibid., 165.

25. Twain, *Following the Equator*, 137.

26. *Electrical World and Engineer* 18, 1891, 39.

27. Shiman, "Explaining the Collapse of the British Electricity Supply Industry in the 1880s," 320–321, 326; Byatt, *The British Electrical Industry, 1875–1914*, 23.

28. Rose, *Cities of Heat and Light*, 18–20.

29. Whipple, *Municipal Lighting*, 56.

30. Sharpe, "Brief Outline of the History of Electric Illumination in the District of Columbia," 204–205.

31. Schivelbusch, *Disenchanted Night*, 99–106.

32. Armengaud, Armengaud, and Cianchetta, *Nightscapes: Paisajes Nocturnos*, 72–73.

33. Schivelbusch, *Disenchanted Night*, 97–98.

34. Bedarida and Sutcliffe, "The Street in the Structure and Life of the City," 22–26.

35. Ibid., 31.

36. Ibid., 27–28.

37. Ibid., 31–35.

38. Wrightington, "Street Lighting with Gas in Europe," 534.

39. Cavling, *Fra Amerika*, 322–324.

40. Onuf, "Liberty, Development, and Union"; Johnson, "Towards a National Landscape."

41. On US city plans, see Reps, *The Making of Urban America.*

42. This paragraph based on Nye, *American Technological Sublime*, 45–108.

43. Sears, *Sacred Places*, 3–6.

44. Otter, *The Victorian Eye*, 185.

45. Cited in ibid., 185.

46. Gustavus Hartridge, "The Electric Light and Its Effects upon the Eyes," *British Medical Journal* 1, no. 1625 (February 20, 1892): 382–383.

47. Adams, *Our American Cousins*, 202–203.

48. Pratt, *The Electric City*, 69–77.

49. Inwood, *City of Cities*, 288.

50. Cited in Parsons, *The Early Days of the Power Station Industry*, 189.

51. Byatt, *The British Electrical Industry, 1875–1914*, 99.

52. Schott, "Empowering European Cities," 169–173; Byatt, *The British Electrical Industry, 1875–1914*, 26, 27–28.

53. Schott, "Empowering European Cities," 176.

54. Ibid., 175; Klingenberg, "Electricity Supply of Large Cities," 138–139.

55. Parsons, *The Early Days of the Power Station Industry*, 198.

56. Inwood, *City of Cities*, 286.

57. Hughes, *Networks of Power*, 227–228.

58. Haskell, "Architecture, the Bright Lights," 55–56.

59. Chesterton, *What I saw in America*, 33.

60. Nye, *America's Assembly Line*, 53.

61. McShane, *Down the Asphalt Path*, 57–80.

62. Rae and Williams, "Creating Demands for Electricity," 751–760.

63. Inwood, *City of Cities*, 280.

64. Nela Park Collection, box 1, folder 7, "European Diary," Smithsonian Institution Archives, Washington, DC.

Chapter 4: Moonlight Towers

1. O'Dea, The Social History of Lighting, 100, citing *Gentleman's Magazine*, February 1763.

2. "Lighting the City by Towers," *Hazard's Register of Pennsylvania*, 55.

3. Nye, *Electrifying America*, 29.

4. Schivelbusch, *Disenchanted Night*, 124–127.

5. *Proceedings of the National Electric Light Association, 1886*, 33.

6. "Electrical Miracles of Tomorrow," *Literary Digest*, 24, http://www.unz.org/Pub/LiteraryDigest-1925jul18-00024.

7. "Edison's Prophecy," *Literary Digest*, 966–968.

8. Wrege, "J. W. Starr, Cincinnati's Forgotten Genius," 104.

9. "Lights for Public Spaces: Plans of the Brush Company in Fifth Avenue," *New York Times*, May 5, 1881, 8.

10. Preece, "Our Lights as Others See Them," 264.

11. "Electric Lights in a Cluster," *New York Times*, September 4, 1880, 3. See also Hammer Papers, box 25, folder 7, unidentified press clipping on a proposal to try

tower lighting in Holyoke, Massachusetts, Smithsonian Institution Archives, Washington, DC.

12. Whipple, *Municipal Lighting*, 41, 47.

13. Hammond, *Men and Volts*, 146.

14. On arc lights and platform lighting, see also Jakle, *City Lights*, 38–58.

15. Israel, *Edison: A Life of Invention*, 187.

16. Helm, *History of Wabash County*, 240; Nye, *Electrifying America*, 3.

17. Nye, *American Technological Sublime*, 143–172.

18. Tocco, "The Night They Turned the Lights on in Wabash," 359.

19. Hammond, *Men and Volts*, 71.

20. "Jenney Electric Light Company," Indianapolis, catalog, ca. 1887, Library, Museum of American History, Smithsonian Institution, Washington, DC.

21. Whipple, *Municipal Lighting*, 98.

22. Cited in Jakle, *City Lights*, 52.

23. Examples from Whipple, *Municipal Lighting*, passim.

24. Ibid., 42, 53.

25. Jakle, *City Lights*, 48.

26. Kaempffert, *Ornamental Street-Lighting,* 25.

27. "Electric Lighting in the City," *Engineering,* 31.

28. *Proceedings of the National Electric Light Association, 1886*, 159, 174; "Report on Detroit," *The Electrical World*, May 5, 1888, 233.

29. "Tower Electric Lighting," *Honolulu Daily Bulletin*, October 20, 1885, 3.

30. Whipple, *Municipal Lighting*, 157.

31. Cited in *Detroit Post*, December 17, 1884.

32. Mark Twain, *Mark Twain's Notebooks and Journals*, Volume III, ed. Robert Park Browning, Michael B. Frank, and Lin Salama (Berkeley: University of California Press, 1979), 81.

33. *Proceedings of the National Electric Light Association, 1885*, 170.

34. *Detroit Journal*, July 13, 1884.

35. "Streets Lit by Towers: The Detroit Electric Lighting System," *New York Times*, August 20, 1885, 8.

36. "Chicago's Electric Blaze," *New York Times*, January 1, 1886, 1.

37. *Proceedings of the National Electric Light Association, 1886*, 67–68.

38. "Electrical News and Notes," *Electrician and Electrical Engineer*, 389.

39. "The Detroit Electric Light Convention," *Electrician and Electrical Engineer*, 388.

40. "Detroit Meeting of the National Electric Light Association," *Electrician and Electrical Engineer*, 361.

41. *Proceedings of the National Electric Light Association, 1886*, 3.

42. *Omaha Daily Bee*, July 13, 1887, 6. On Allegheny, see *Wheeling Daily Intelligencer*, May 1, 1890, 3.

43. Dow, "Public Lighting in Relation to Public Ownership and Operation," 93–116.

44. *Proceedings of the National Electric Light Association, 1885*, 175.

45. Freeberg, *The Age of Edison,* 67–69.

46. "Tower System for Electric Arc Lighting," *American Architect and Building News*, 11, 227.

47. Ibid.

48. All quotations from "The Tower System," *Chicago Daily Tribune*, July 25, 1881, 3.

49. *Electrical World*, April 26, 1884, 139.

50. "Pillar of Fire," *Omaha Daily Bee*, 6.

51. "Tower System for Electric Arc Lighting," *American Architect and Building News*, 227.

52. *Los Angeles Herald*, October 16, 1881, 1.

53. Isenstadt, "Los Angeles," 51.

54. Whipple, *Municipal Lighting*, 18.

55. Huebinger, *First Album of the City of Davenport, Iowa*, 6.

56. "Jenney Electric Light Company," Indianapolis, catalog, ca. 1887, Trade Catalog Collections, Museum of American History, Smithsonian Institution, Washington, DC.

57. *American Architect and Building News*, November 20, 1880, 8, 256.

58. "Proposed Electric Light Tower at New Orleans," *Scientific American*, 11, 159.

59. Ibid, 159.

60. Jakle, *City Lights*, 47.

61. Twain, *Life on the Mississippi*, 427, 520.

62. Kendall, *History of New Orleans*, chapter 29.

63. Whipple, *Municipal Lighting*, 51.

64. Testimonial letter in catalog, "The Fort Wayne 'Jenney' Electric Light Co.," Fort Wayne, Indiana, n.d. (ca. 1888), Warshaw Collection, box 15, 33, Smithsonian Institution Archives, Washington, DC.

65. Beaumont, *Night Walking*, 343.

66. Homann, *Night Vision.*

67. Milder, *Reimagining Thoreau*, 109.

68. Thoreau, "Night and Moonlight," 579.

69. Hawthorne, *The Scarlet Letter*, 149.

70. Whipple, *Municipal Lighting*, 62.

71. Ibid., 60–61.

72. Ibid., 67, 62.

73. Ibid., 163.

74. Ibid., 62.

75. Ibid., 167.

76. "The Electric Masts in Cleveland," *New York Times*, July 12, 1886, 2.

77. Preece, "Our Lights as Others See Them," 264.

78. Austin's sixteen towers, still in operation, are registered as a historic landmark. Moore and Strand, "Preservation Study of the Moonlight Towers, Austin, Texas."

79. "Municipal Lighting," *Electrical World*, July 7, 1888, 2.

80. Whipple, *Municipal Lighting*, 74.

81. Atless, *Nocturne*, 95.

82. Cited in ibid.

83. "Lunar-Resonant Streetlights," Civil Twilight Design Collective, September 6, 2016, accessed May 10, 2017, http://www.civiltwilightcollective.com/lunar1.htm.

Chapter 5: Spectacles and Expositions

1. "A Grand Display," *Boston Globe*, October 15, 1880, 1.

2. "Yesterday at Yorktown: Opening Day of the Centennial," *Washington Post*, October 19, 1881, 1.

3. The Brush Company provided its equipment to be used free of charge, except for the cost of transportation from New York and salary of a supervising engineer. "Preparing for the Carnival," *Washington Post*, September 15, 1881, 4.

4. "Lighting of Cheyenne, Wyoming," Hammond Papers, General Electric Library, B-174–176, Schenectady, NY.

5. Letter, Thomas A. Edison to Steele MacKaye, 1880, Edison Papers, Edison National Historic Site, West Orange, NJ.

6. Marer, *David Belasco*, 78–82; Henderson, *Theater in America*, 230–235.

7. "Light on the Circus," *Washington Post*, April 29, 1879. See also Sharpe, "Brief Outline of the History of Electric Illumination in the District of Columbia," 192.

8. *Washington Evening Star*, March 5, 1881, 1.

9. Sharpe, "Brief Outline of the History of Electric Illumination in the District of Columbia," 202.

10. Ibid., 200–203.

11. "Veiled Prophet," *Harper's*, 667.

12. Mohr, *Excursion through America*, 293–295.

13. "Street Illuminations of St. Louis," *Frank Leslie's Popular Magazine*, 25. See also Spencer, *The St. Louis Veiled Prophet Celebration*.

14. Half-page advertisement, *Fort Worth Gazette*, September 12, 1893.

15. Cox, *St. Louis through a Camera*, 83.

16. "Electricity in the West," *Frank Leslie's Popular Monthly*.

17. Beauchamp, "The Mystery of St. Louis's Veiled Prophet."

18. "The Veiled Prophet," *Harper's Weekly*.

19. "The Presidential Tour," *Harper's Weekly*, 143.

20. Spencer, *The St. Louis Veiled Prophet Celebration*, 75–77.

21. Ibid., 78.

22. "Triumphant Industries: The General Electric Company," *Forum*, February 18, 1893, in Warshaw Collection, electricity series, box 15, Smithsonian Institution Archives, Washington, DC.

23. "The Milwaukee Carnival," *Street Railway Journal*, August 4, 1900, 705.

24. Glassberg, "Public Ritual and Cultural Hierarchy," 421–448.

25. Hunter, *Steam Power*, 295–299.

26. Neumann, *Architecture of the Night*, 10.

27. Nye, *Electrifying America*, passim.

28. Vegas and Mileto, "World's Fairs," 177.

29. Dreiser, *Newspaper Days*, 309–310.

30. Adams, *The Education of Henry* Adams, 1067–1068.

31. Beltran and Carré, *La fée et la servante*, 64–72.

32. Moncel and Preece, *Incandescent Lights, with Particular Reference to the Edison Lamps at the Paris Exhibition*, 49. Copy in the Hammer Collection, box 44, Smithsonian Institution Archives, Washington, DC.

33. Ibid., 47–48.

34. Siemens had exhibited a smaller electric tram in 1879 at the electrical fair in Berlin, but it did not carry the general public.

35. *Louisville Courier-Journal*, July 4, 1883, Hammond Papers, General Electric Library, L 5410–5414, Schenectady, NY.

36. Stieringer, "The Evolution of Exposition Lighting," 187.

37. "The Electric Age," *Milwaukee Sentinel,* September 7, 1884; "The Electrical Exposition," *Harper's Weekly.*

38. *Journal of the Franklin Institute,* May 1886, 121.

39. Gibson, "The International Electrical Exhibition of 1884 and the National Conference of Electricians," 44, 85.

40. "The International Electrical Exposition, Philadelphia," *Scientific American,* 192.

41. *Boston Herald,* cited in "New Orleans Exhibition Building," *Southern Planter,* August 1884, 404; "The Cotton Centennial Exhibition at New Orleans," *American Architect and Building News.*

42. Stieringer, "The Evolution of Exposition Lighting," 187.

43. *Official Guide of the Ohio Valley and Central States Exposition,* 15.

44. *Cincinnati Commercial Gazette,* June 10, 1888, 4, Hammer Papers, box 42, folder 4, Smithsonian Institution Archives, Washington, DC.

45. Clipping book, Hammer Papers, box 47, Smithsonian Institution Archives, Washington, DC.

46. *Proceedings of National Electric Light Association,* Appendix B (1903): 23.

47. See Stieringer, "The Evolution of Exposition Lighting," 187.

48. Williams, "Decorative and Sign Lighting," 84.

49. *Proceedings of National Electric Light Association* (1903): 24.

50. Dickerson, "Spectacular Lighting," 485.

51. "Next to Edison Stands Stieringer," *Los Angeles Times,* December 21, 1902, C10.

52. Rydell, *All the World's a Fair,* 40.

53. Martin and Stieringer, "On the Electric Lighting of the World's Fair," 189.

54. Ibid., 208.

55. "Stars Made to Pale," *Daily Inter Ocean,* May 14, 1893.

56. "World's Fair Doings," *Daily Inter Ocean,* December 8, 1891, 4.

57. "Wonderland in Electric Building," *Current Literature,* 21.

58. Barrett, *Electricity at the Columbian Exposition,* 16–19.

59. Caption accompanying reproduction in *The Columbian Gallery,* 86.

60. Levinson, Samuels, Vandersee, and Winner, *The Letters of Henry Adams,* Volume 4, 132.

61. Haynes, *History of the Trans-Mississippi and International Exposition of 1898.*

62. Stieringer, "The Evolution of Exposition Lighting," 188.

63. "Shown by Electric Light," *Omaha Daily Bee,* June 2, 1898, 1.

64. Cited in Haynes, *History of the Trans-Mississippi and International Exposition of 1898*, 141. See also John A. Wakefield, "A History of the Trans-Mississippi and International Exposition," typescript, May 20, 1903, Omaha Public Library.

65. Quotation in Williams, "Decorative and Sign Lighting," 85.

66. Mandell, *Paris, 1900*, 112–113.

67. Ibid., 114–115.

68. Stieringer, "The Evolution of Exposition Lighting," 189.

69. Ibid., 189.

70. Ibid., 189–190.

71. Brush, "Electrical Illumination at the Pan-American Exposition," 20944.

72. Neumann, *Architecture of the Night*, 88.

73. Rydell, *All the World's a Fair*, 131.

74. Turner, "The Color Scheme," 21.

75. *The Wonders of the World's Fair*, plate 48 (unpaged).

76. Turner, "The Color Scheme," 21.

77. Bell, "Elements of Illumination," 221.

78. Mabel E. Barnes, "Peeps at the Pan-American," manuscript, 3:144, Erie County Historical Society, Erie, PA. See also Barry, *The Grandeurs of the Exposition*, centerfold.

79. Barry, *The Grandeurs of the Exposition*, centerfold.

80. Brush, "Electrical Illumination at the Pan-American Exposition," 20943.

81. Grant, "Notes on the Pan-American Exposition," 454.

82. "Edison at the Pan-American Exposition," 103. See also "Edison, Stieringer, Rustin, Meeting of the Wizard," *Omaha World Herald*, August 22, 1901.

83. Barnes, "Peeps at the Pan-American," 3:146–147.

84. Huhtamo, "The Sky Is Not the Limit," 329; Nye, *Electrifying America*, 38–39.

85. Cited in "Topics of the Times," *New York Times*, September 16, 1901, 6.

Chapter 6: Commercial Landscape

1. A minority preferred a third alternative, based on organic architectural conceptions. The central spokespeople for this point of view were Louis Sullivan and Frank Lloyd Wright. Sullivan saw the architecture of the Columbian Exposition as a falsification and anticipated modernism with his dictum "form follows function." From his perspective, neither the gothic nor the Beaux-Arts style was suitable for expositions or skyscrapers, but this was a minority perspective before 1920.

2. Fisher, *Still the New World*, 47–50; Nye, *America as Second Creation*, 21–42.

3. Larwood, *The History of Signboards, from the Earliest Times to the Present Day*, 15, 19.

4. Ibid., 20.

5. Ibid., 31, 38.

6. Bressani, "Paris," 28.

7. Mills, "The Development of Electric Sign Lighting," 364.

8. Brochure, D. A. Baker Company, Warshaw Collection, advertising series, box 6, folder 2, Smithsonian Institution Archives, Washington, DC. Similarly, a Welsbach fixture designed for shop windows could be used with either gas or electricity. Brochure, Warshaw Collection, lighting series, box 4, folder 3, Smithsonian Institution Archives, Washington, DC.

9. Bressani, "Paris," 31–32.

10. Shultz, "Legislating Morality," 37.

11. Laird, *Advertising Progress*, 63.

12. "Poster Advertising," *Advertising Experience: A Magazine for American Advertisers*, December 1896, 13, Warshaw Collection, advertising series, box 5, folder 10, Smithsonian Institution Archives, Washington, DC.

13. "Outdoor Displays in Detroit," *Billposter: Display Advertising* 8, no. 1 (March 1903): 8, Warshaw Collection, advertising series, box 5, folder 16, Smithsonian Institution Archives, Washington, DC.

14. Baker, "Public Sites versus Public Sights," 1191.

15. Presbury, *The History and Development of Advertising*, 506; Burrows and Wallace, *Gotham*, 645.

16. Bach, "To Light up Philadelphia," 325.

17. McCabe, *New York by Gaslight*, 252–253.

18. Smith, *Sunshine and Shadow in New York*, 27, 707, 709.

19. Heap, *Sexual and Racial Encounters in American Nightlife, 1885–1940*, 29–45.

20. Ibid., passim.

21. Sharpe, *New York Nocturne*, 145–146.

22. Riis, *How the Other Half Lives*.

23. Laird, *Advertising Progress*, 157.

24. Archer, *America Today*, 29.

25. Hammer Papers, box 19, folder 6.

26. McAllister, "Possibilities of Sign and Decorative Lighting," 320–321, 329.

27. Spaulding, "Display Lighting, Signs, and Decorative Light," 311.

28. Williams, "Decorative and Sign Lighting," 9–11.

29. Gilchrist, "Electric Signs," 318–319.

30. "Darkened Broadway Just a War Reminder," *New York Times*, December 16, 1917, 4.

31. Lippincott, *Outdoor Advertising*, 97.

32. Horace Zollars, "Talking Lights of Chicago; Marvels of Electric Signs," *Chicago Daily Tribune*, May 16, 1909, E3.

33. Davis, *The Great Streets of the World*, 26.

34. Nye, *Consuming Power*, 160.

35. Nasaw, *Going Out*, 4–5.

36. Heinze, *Adapting to Abundance*, 23.

37. "Completion of the Tall Tower," *New York Times*, October 14, 1891, 9.

38. "A Pyramid of Fire, Complete Illumination of the Madison Square Garden," *New York Times*, November 2, 1891, 5; "A Huge Weather Vane, Diana in Copper," *New York Times*, September 29, 1891, 9.

39. Mills, "The Development of Electric Sign Lighting," 370–371.

40. Traub, *The Devil's Playground*, 50–51. See also Leach, *Land of Desire*, 47–48.

41. "The Lights of Broadway," *Literary Digest*.

42. Ziegler, "The Living Electric Sign," 74–75.

43. Irwin Ellis, "Dayton Genius Again to the Front; Electric Sign Its Latest Wonder," *Chicago Daily Tribune*, April 24, 1910, E6; Presbury, *The History and Development of Advertising*, 509.

44. *Edison Light*, March 1903, 7.

45. Nye, *American Technological Sublime*, 179.

46. Nye, *Electrifying America*, 56.

47. "Blaze of Glory," *Los Angeles Times*, April 30, 1905, 3.

48. Kaempffert, *Ornamental Street-Lighting*, 17.

49. Nye, *Electrifying America*, 52.

50. "Advertising the Church," *Literary Digest*.

51. National Electric Light Association, box 3, folder 1, "Let's See the Church Windows," *Simulator*, December 1919, 11.

52. Baker, "Public Sites versus Public Sights," 1208.

53. "Grand Christmas Display," *Boston Daily Globe*, December 15, 1896, 12.

54. Hannah, *Electricity before Nationalisation*, 3.

55. Pincus, "Common Errors in Park Construction," 461.

56. "Some Features of Cost Keeping and Accounting at Willow Grove Park, Philadelphia," *Electric Railway Journal*, 370.

57. Kasson, *Amusing the Million*, 49–50.

58. Pincus, "Common Errors in Park Construction," 461.

59. Burne-Jones, *Dollars and Democracy*, 56, 60.

60. Archer, *America Today*, 45–46.

61. "Electric Signs for London," *New York Times*, July 14, 1912, C4.

62. "Curb on Electric Signs," *New York Times*, March 8, 1914, C4.

63. "Beginning and Development of Illuminating Engineering, Information Obtained from W. D'Arcy Ryan, Oct 27 and 28, 1925." Hammond Papers, General Electric Historical File, 1648–1649, Schenectady, NY.

64. Harrison, "The Importance of Artificial Light to Architecture," 475–478.

65. Koron, *The American Skyscraper*, 276.

66. Cather, "Behind the Singer Tower."

67. Fenske, *The Skyscraper and the City*, 3.

68. Talbot, *Electrical Wonders of the World*, 1:535–538; "Fifty-Five Story Building Opens on a Flash," *New York Times*, April 1913.

69. Luckiesh, *Artificial Light*, 300–301.

70. Fenske, *The Skyscraper and the City*, 218.

71. Ibid., 285.

72. Luckiesh, *Artificial Light*, 302.

73. Ibid., 30.

74. Fenske, *The Skyscraper and the City*, 266–270.

75. Ibid., 283.

76. Fenske, *The Skyscraper and the City*, 62, 64.

77. Nye, *American Technological Sublime*, 103–108.

Chapter 7: City Beautiful

1. Baldwin, *In the Watches of the Night*, 181.

2. Robinson, *The Improvement of Towns and Cities*, 86.

3. "Outdoor Advertising in France," *Journal of the Society of Arts*, 1058.

4. "Curb on Electric Signs," *New York Times*, March 8, 1914, C4.

5. Gilbert, *Perfect Cities*, 90–94.

6. "The Desecration of Scenery," *Chicago Daily Tribune*, October 7, 1900, 40.

7. "Beautify the Streets," *Current Literature*, 513.

8. "Eyesores to Civic Beauty," *Los Angeles Times*, April 4, 1906, sec. II, 10.

9. W. C. Taylor, "Electric Curiosities," *New York Times*, September 11, 1910, 10.

10. "All Now in Accord on Billboard Rules," *New York Times*, May 11, 1914, 7.

11. Tauranac, *The Empire State Building*, 50–59.

12. Jacobs, "Downtown Is for People," 183.

13. Walker, "The City of the Future—A Prophecy."

14. Wilson, *The City Beautiful Movement*, 92; Boyer, *Urban Masses and Moral Order in America,* 1820–1920, 253.

15. Roemer, *The Obsolete Necessity*. See also Segal, "Edward Bellamy and Technology," 101–116.

16. Segal, "The Technological Utopians," 119.

17. Peterson, *The Birth of City Planning in the United States, 1840–1917*, 151.

18. Missal, *Seaway to the Future*, 123.

19. Cited in ibid., 123.

20. Ibid., 135, 141. See also Haskin, *The Panama* Canal, 165–174.

21. Missal, *Seaway to the Future*, 188–193.

22. "Beautify the Streets," *Current Literature*, 513.

23. Wilson, *The City Beautiful Movement*, 61.

24. Ibid., 1.

25. "Outdoor Advertising in Germany," *Journal of the Society of Arts*, 606–607.

26. Robinson, *The Improvement of Towns and Cities.*

27. Burnham and Bennett, *Plan of Chicago.*

28. Kahn, *Imperial San Francisco*, 90; Burnham, *Report on a Plan for San Francisco.*

29. Robinson, *The Improvement of Towns and Cities*, 6–7, 134, 139.

30. Ibid., 51–52. See also Eshelman, "Modern Streetlighting."

31. Robinson, *The Improvement of Towns and Cities*, 79–80.

32. Robinson, "Improvement in City Life," 773.

33. Rae and Williams, "Creating Demands for Electricity," 751.

34. Robinson, "Improvement in City Life," 771–772.

35. Layton, *The Revolt of the Engineers.* A few engineers followed the lead of Charles Steinmetz, who was an evolutionary socialist. Kline, *Steinmetz.*

36. Cited in Tichi, *Shifting Gears*, 184.

37. S. E. Doane, "A Civic Duty for Engineers" (speech, Engineering Department, National Electric Lamp Association, Cleveland, OH, December 5, 1916), 16–17, 24, 30–31, 37, 14.

38. Bolton, "The Great Awakening of the Night," 42–43.

39. Rossell, "Compelling Visions," 45, 47.

40. Hammond Papers, interview with W. D'Arcy Ryan, 1041–1042, General Electric Library, Schenectady, NY.

41. Bolton, "The Great Awakening of the Night," 43.

42. Hughes, *Networks of Power*, 222–225.

43. Keating, *Lamps for a Brighter America*, 72, 78.

44. Hammond Papers, interview with Ryan, 1042.

45. Gilbert, *Whose Fair?*, 38.

46. Cited in Rydell, *All the World's a Fair*, 160–161.

47. Adas, *Machines as the Measure of Men*, 339–340.

48. During preparations for the fair, Saint Louis discussed construction of a harmonious public buildings group in its center and planned a boulevard connecting two large parks. Sandweiss, *St Louis*, 193–195.

49. Peterson, *The Birth of City Planning in the United States, 1840–1917*, 147–148.

50. "The Model City at St. Louis," *Charities*, 126–127.

51. Francis, *The Universal Exposition of 1904*, 416.

52. Ibid., 420.

53. "Meet Me in St. Louis, Louis," lyrics by Andrew B. Sterling, music by Kerry Mills, Victor recording, 2850, 1904.

54. Everett, *The Book of the Fair, the Greatest Exposition the World Has Ever Seen,* 201–203.

55. Adams, *The Education of Henry Adams*, 1146.

56. Luckiesh, *Artificial Light*, 300.

57. "Official Minutes of the Hudson-Fulton Celebration Commission, Together with the Minutes of Its Predecessor, the Hudson Tercentenary Joint Committee," May 5, 1909, 1085, accessed May 14, 2017, http://www.archive.org/stream/officialminuteso00hudsa/officialminuteso00hudsa_djvu.txt. On the celebration, see Nye, *American Technological Sublime*, 147–172.

58. Ibid., May 18, 1909, 1099; ibid., June 2, 1909, 1157.

59. See ibid., June 30, 1909. This meeting indicated that the electrical companies were willing to illuminate twenty-two parks for $10,000, and that the line of the parade march was to have additional intensive lighting (ibid., 1209).

60. Ibid., October 13, 1909, 1561.

61. Hall, *The Hudson Fulton Celebration*, 122.

62. Ibid.

63. "Spectacular Electric Illumination during the Hudson-Fulton Celebration," *Electrical World,* 290. See also the *Edison Monthly*, October 1909, 120–133.

64. Hall, *The Hudson Fulton Celebration*, 120–121.

65. Ibid., 121.

66. Calendar of events, Hudson-Fulton Celebration, inserted in ibid., 19.

67. Nye, *American Technological Sublime*, 163–165.

68. McLaren and Ryan, "Coloring a City," 346.

69. Ryan, "The Illumination of the Panama-Pacific Exposition," *General Electric Review*, 580.

70. Ibid.

71. Rossell, "Compelling Vision," 100.

72. Ryan, "The Illumination of the Panama-Pacific Exposition," 580–581.

73. Ackley, *San Francisco's Jewel City*, 137.

74. Nye, *Electrifying America*, 3.

75. Todd, *The Story of the Exposition*, vol. 2, 345.

76. Luckiesh, *Artificial Light*, 309.

77. Ackley, *San Francisco's Jewel City*, 132.

78. Brechin, "Sailing to Byzantium," 111.

79. Ackley, *San Francisco's Jewel City*, 131.

80. Ryan, "The Illumination of the Panama-Pacific Exposition," 582.

81. "Grand Prize for Lighting: Welsbach Company Gets Highest Award at the Panama Pacific," *Boston Daily Globe*, July 5, 1915, 12.

82. Cited in Bolton, "The Great Awakening of the Night," 46.

83. Rossell, "Compelling Vision," 101, 109–110.

84. *Journal of Electricity, Power, and Gas* 37, no. 11 (1916), accessed May 15, 2017, http://archive.org/stream/journalofele371231916paci/journalofele371231916paci_djvu.txt.

85. Barrett, *Our Wondrous Trip*.

Chapter 8: Light as Political Spectacle

1. "Lighting the Capitol," *Washington Post*, November 20, 1878, 1.

2. "A New President," *Boston Globe*, March 5, 1881, 1.

3. "Evening Scenes: The Inauguration Ball," *Boston Daily Globe*, March 5, 1881, 1.

4. Benjamin Harrison, "Remarks during Decoration Day Ceremonies at Laurel Hill Cemetery in Philadelphia," May 30, 1891, accessed May 15, 2017, http://www.presidency.ucsb.edu/ws/index.php?pid=76212&st=electric&st1=light.

5. Schlereth, "Columbia, Columbus, and Columbianism," 121–126.

6. Jakle, *City Lights*, 129.

7. Perry Belmont et al., "Some Suggestions for the Exterior Decoration of Dwelling Houses, Clubs, and Hotels, by the Committee on Art for the Columbian Celebration," New York, October 8, 9, 10, 11, 12, and 13, 1892, pamphlet, 1892, 16, copy in New York Historical Society.

8. Arthur Williams, "Lighting New York's Streets" [first presented as a radio talk and then published as a booklet], GE Historical file, Publicity Department, Hammond Papers L6616, 4.

9. Carletta, "The Triumph of American Spectacle," 34.

10. Ibid., 19–40. See also Marvin, *When Old Technologies Were New*, 107–109; Nardis, *Wonder Shows*, 55.

11. Trouillot, "Good Day Columbus," 1–24.

12. Schlereth, "Columbia, Columbus, and Columbianism," 127.

13. "Dewey, Dewey, Rah, Rah, Rah," *Minneapolis Journal*, September 26, 1899, 1.

14. Fenske, *The Skyscraper and the City*, 218.

15. "Sky Seemed Afire," *Baltimore Sun*, September 30, 1899, 2.

16. "Illumination of New York, City, and Harbor Aflame with Fireworks in Dewey's Honor," *New Haven Evening Register*, September 30, 1899, 3.

17. Cited in *Literary Digest*, October 7, 1899, 421.

18. Bangs, "Dewey's Return," 935.

19. "Dewey's Welcome," *Helena Independent*, September 26, 1899, 7.

20. Barnes, "The Story of Dewey's Welcome Home," 312.

21. Marvin, *When Old Technologies Were New*, 158–159.

22. Anderson, "Brief Outline of Electric Sign History," 18.

23. Marvin, *When Old Technologies Were New*, 167.

24. Ibid., 186.

25. Jakle, *City Lights*, 124.

26. Anderson, "Brief Outline of Electric Sign History," 19.

27. "Inaugural Was Most Splendid in History," *Aberdeen Daily American*, March 5, 1909, 1.

28. "Blaze of Fireworks Marks Celebration," *State* (Columbia, SC), March 5, 1913, 14.

29. "60,000 in Park Hear Community Chorus," *New York Times*, September 14, 1916, 5.

30. Massey, "Organic Architecture and Direct Democracy," 595–596.

31. Cited in ibid., 598.

32. Ibid., 602.

33. Wainwright, *History of the Philadelphia Electric Company, 1881–1961*, 131–133.

34. Millar, "Wartime Lighting Economy," 332.

35. "Darkened Broadway Just a War Reminder," *New York Times*, December 16, 1917, 4.

36. Schlereth, *Victorian America*, 169.

37. Marvin, *When Old Technologies Were New*, 167.

38. Plotnick, "At the Interface."

39. Fenske, *The Skyscraper and the City*, 218.

40. Moore, *Empire on Display*, 68, 97.

41. Nathan-Garner, *Insider's Guide to Houston*, 32.

42. "About the Centennial of the Panama-Pacific International Exposition," accessed May 16, 2017, http://www.ppie100.org/about/.

43. Fishback. "Chronicler's Report for 1915," 222-223.

44. Coopman, *The History of Rock Island County*, chapter 4, unpaginated.

45. "Lights to Spell Out High Ideals on 4th," *New York Times*, July 2, 1916, S4.

46. "Unveiling of Large Electric Flag Crowning Event," *Signs of the Times*, 17.

47. Nye, *American Technological Sublime*, 264–266.

48. Luckiesh, *Artificial Light*, 304.

49. "Wilson Feted," *Morning Oregonian*, November 30, 1916, 2.

50. "Liberty's New Halo Will Shine Tonight," *New York Times*, December 2, 1916, 4.

51. Blumberg, "A National Monument Emerges," 211–212. See also illustrations, 245, 247.

52. "Macon to Have Electrical Celebration Lasting a Week," *Macon Daily Telegraph*, September 1, 1916, 10.

53. "Plan Electrical Week," *Philadelphia Inquirer*, September 8, 1916, 10.

54. "P. O. Electric Winner," *Washington Post*, December 6, 1916, 4.

55. "Light Is to Flood City," *Washington Post*, February 12, 1917, 10.

56. Van Schaack, "The Division of Pictorial Publicity in World War I," 32–33.

57. Creel, *How We Advertised America*, 133.

58. Herbert S. Houston, "Advertising the New Liberty Bonds," *New York Times*, September 20, 1917, 12.

59. "Outdoor Advertising Men to Give More War Aid to the Government," *Chicago Daily Tribune*, September 11, 1918.

60. "Outdoor Art Association Makes War on Billboards," *Chicago Daily Tribune*, June 7, 1900, 7.

61. Starr and Hayman. *Signs and Wonders*, 66.

62. Nye, *Electrifying America*, 61.

63. "Plumed Parade Stirs Crowds on Fifth Avenue," *New York Times*, May 7, 1919, 3.

64. Brown, "Annihilating Time and Space," 170–201.

65. "France Rejoices Over End of War," *Philadelphia Inquirer*, June 30, 1919, 14.

66. Luckiesh, *Artificial Light*, 305.

67. "Pageantry for Returning Heroes," *Literary Digest*, 26–27.

68. Luckiesh, *Artificial Light*, 307.

69. For a more detailed account of the International Festival of Peace, see Nye, *American Technological Sublime*, 167–171.

70. Luckiesh, *Artificial Light*, 307.

71. "City Bathed in Light," *Washington Post*, August 13, 1919, 7.

72. Barnaby, *Light Touches*, 116–118.

CHAPTER 9: CONCLUSIONS

1. Beaumont, *Night Walking*, 342–344.

2. Cited in ibid., 344.

3. Domosh, *Invented Cities*, 99–126.

4. Leach, *Land of Desire*.

5. Nye, "Implementing a New Energy Regime in Housing," 36–38.

6. Nye, *America's Assembly Line*, 30–33, 56–57.

7. Jakle, *City Lights*, 77.

8. W. D'Arcy Ryan, "Street Lighting," Hammond File, L-1908, General Electric Company Library, Schenectady, NY, 12.

9. Ibid., 13.

10. Jakle, *City Lights*, 94.

11. *American City* 16 (1916): 70.

12. Binder and Reimers, *All the Nations under Heaven*, 96–148.

13. Jacobs, *Life and Death of American Cities*, 152–171.

14. On the US social construction of electricity, see Nye, *Electrifying America*, 138–176.

15. Dreiser, *Newspaper Days*, 310.
16. Luckiesh, *Artificial Light*, 302–303.
17. Cited in Buck-Morss, *The Dialectics of Seeing*, 309.

Glossary

This glossary features the more common forms of gas and electric illumination. There were well over a hundred different gas burners and electric lights on the market in the 1890s, so the following is only a brief introduction to a complex subject. (Images are not to the same scale.)

1. Street lights burning coal gas (1807–1920) provided ten to fifteen candlepower, yet also dirtied the glass that protected the flame from the wind. The light depicted was used in New York City, circa 1870. Each unit needed to be lighted individually, but all the lights on one line could be extinguished together by shutting off the flow of gas from the central plant. In later years they could be lighted automatically using an electric spark. *Source:* nypl.digitalcollections.510d47e1-0fe0-a3d9-e040-e00a18064a99.001.w.jpg.

2. The arc light was the most common form of electric lighting between 1875 and 1910. The DC arc produced two thousand candlepower by jumping a strong current between two carbon rods. Its brightness could not be adjusted. The unit depicted has two sets of rods. When one pair had burned down, the other would automatically come on. The upper rod produced more than 80 percent of the light. It was positively charged and became much hotter than the lower rod. The AC arc light spread after 1893. When further improved, it was enclosed and spread light more widely. In its last years, the rods only needed replacement after burning six hundred hours. It was gradually replaced after circa 1910 by tungsten incandescent lighting (see page 251, no. 7). *Source:* Photograph from Jenney Company Catalog, 1886, Smithsonian Library, by David E. Nye.

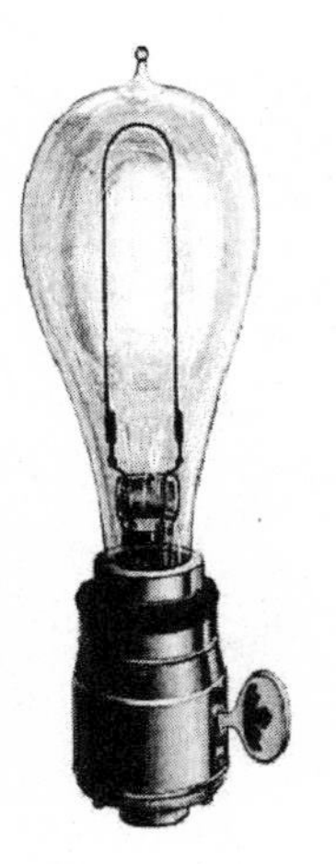

3. Edison's early incandescent light had an enclosed carbon filament, and came in a variety of sizes and strengths. Invented in 1879, it was chiefly utilized indoors, using bulbs of sixteen candlepower. It was less successful as street lighting, but widely adopted for advertising signs and in exposition design. Longer-lasting and more efficient filaments, notably those made from hardened carbon and tungsten, gradually replaced Edison's early light (see page 251, nos. 6 and 7). *Source:* Edison National Historic Site.

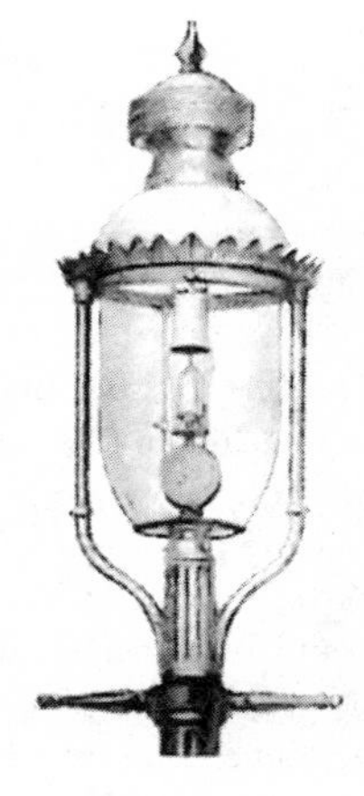

4. The Welsbach gas mantle, which contained oxides of thorium and cerium, surrounded the gas flame. When heated to incandescence, it produced six times more light than burning gas alone. It was widely adopted after 1885, further improved by 1915, and remained common in Europe until the 1920s. *Source: Electrical World,* 1886.

5. The Nernst lamp, invented by Dr. Walther Nernst in 1897, heated to incandescence a mixture of metallic oxides formed into a small rod. A coil of platinum wire preheated this rod because it could not conduct electricity when cold. The lamp produced light close to the daylight spectrum. Acclaimed at the 1900 Paris Exposition, it was manufactured by AEG in Germany and Westinghouse in the United States. It was expensive compared to other forms of lighting, but produced a fine white light. *Source*: *Electrical World,* 1900.

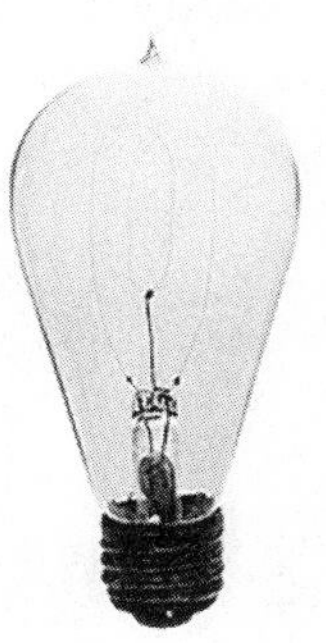

6. The Gem Lamp, compared to the carbon filament bulb, produced one-third more light and lasted 200 hours longer, or 450 hours. Developed by Dr. Willis Whitney, the carbon filament was "metalized" by heating it to thirty-five hundred degrees, thereby removing impurities and creating a hardened graphite shell. In 1905, General Electric sold the Gem Lamp for 25¢, and it was often used in advertising signs. The price declined after that until it went off the market in 1917. *Source:* Courtesy the Schenectady Museum, Hall of History.

7. Alexander Just and Franz Hanaman developed the tungsten incandescent light in Vienna in 1904. At General Electric, William Coolidge improved the filament by making tungsten ductile. His version became available in 1911. Compared to the GEM lamp, it was more efficient, brighter, had an improved light spectrum, and lasted three times as long, or 1,350 hours. In 1912, a sixty-watt ductile tungsten lamp cost 60¢. More powerful tungsten lights outcompeted arc lights and Welsbach gas lamps, and were the basis for lighting the San Francisco Panama Pacific Exposition in 1915. *Source: Electrical World,* 1913.

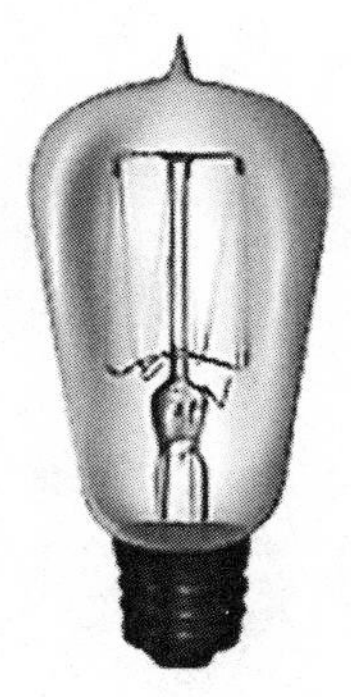

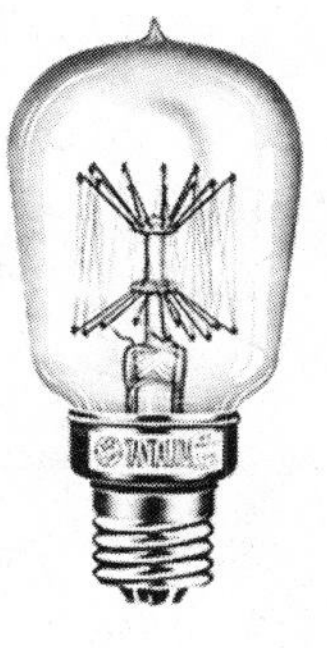

8. Siemens and Halske developed tantalum bulbs, and General Electric and National purchased rights to manufacture them for the US market. Tantalum crystalized, however, when using AC current. It worked well with DC, but that was being phased out by 1906. *Source: Electrical World,* 1915.

9. Baron Welsbach invented the Osmium bulb (not shown). The bulb's filaments were brittle and broke easily. Because osmium is expensive, this bulb sold for more than $1 each. It was never commercially available in the United States.

Bibliography

For readers searching for further sources, this note concentrates on books, though there are many fine articles as well. Exemplary early works on lighting focused on technical issues such as the sequence of inventions, diffusion of lighting systems, or history of specific utilities. These included Bright, *The Electric Lamp Industry* (1949), Passer, *The Electrical Manufacturers* (1953), Keating, *Lamps for a Brighter America* (1954), and Josephson, *Edison* (1959), and on British developments, Byatt, *The British Electrical Industry, 1875–1914* (1979) and Hannah, *Electricity before Nationalisation* (1979). The field became more theoretically sophisticated with the publication of Hughes's *Networks of Power* (1983), which compared developments in the United States, Britain, and Germany; Schivelbusch, *Disenchanted Night: The Industrialization of Light in the Nineteenth Century* (German, 1983, trans. 1988), which looked primarily at Europe; Friedel and Israel, *Edison's Electric Light: Biography of an Invention* (1985); Tarr and Dupuy, *Technology and the Rise of the Networked City in Europe and America* (1988); Beltran and Carré, *La fée et la servante: La societé française face a l'électricité* (1991); and Pratt, *The Electric City: Energy and the Growth of the Chicago Area* (1991). The perception and consumption of electricity were studied by Marvin, *When Old Technologies Were New* (1988) as well as Nye, *Electrifying America: Social Meanings of New Technology (*1990), *American Technological Sublime* (1994), and *Consuming Power: A Social History of American Energies* (1998).

In recent years, books on the history of gas and electric lighting have been so numerous that a chronological list can only be suggestive: Israel's definitive biography *Edison: A Life of Invention* (1998), Schlor's international comparisons in *Nights in the Big City* (1998); Jakle's invaluable US case studies in *City Lights: Illuminating the American Night* (2001), Sharpe's expansive *New York Nocturne: The City after Dark in Literature, Painting, and Photography* (2008), Otter's innovative *The Victorian Eye: A Political History of Light and Vision in Britain, 1800–1910* (2008), Nye's history of blackouts in *When the Lights Went Out* (2010), Werrett's groundbreaking *Fireworks: Pyrotechnic Arts and Sciences in European History* (2010), Koslofsky's indispensable *Evening's Empire: A History of the Night in Early Modern Europe* (2011), Baldwin's US cultural history, *In the Watches of the Night: Life in the Nocturnal City, 1820–1930* (2012), Tomory's *Progressive Enlightenment: The Origins of the Gaslight Industry, 1780–1820* (2012), Carlson's definitive study *Tesla: Inventor of the Electrical Age* (2013), Freeberg's wide-ranging *The Age of Edison: Electric Light and the Invention of Modern America* (2013), Barnaby's new slant on nineteenth-century Britain in *Light Touches: Cultural Practices of Illumination, 1800–1900* (2017), and two groups of essays on the lighting of cities, Isenstadt, Petty, and Neumann, *Cities of Light: Two Centuries of Urban Illumination* (2015), and Meier, Hasenöhrl, Krause, and Pottharst, *Urban Lighting, Light Pollution, and Society* (2015).

The bibliography contains books, magazine stories, and journal articles cited in the text. Newspaper articles and archival sources can be found only in the notes.

"A New Electric Light." *Galaxy: A Magazine of Entertaining Reading* 30, no. 3 (September 1875): 415.

"Advertising the Church." *Literary Digest*, December 27, 1913, 1279.

"Beautify the Streets." *Current Literature* 32, no. 5 (May 1902): 513.

"Cost of Gas and Electricity in New York." *Scientific American* 52, no. 21 (May 23, 1885): 327.

"Detroit Electric Light Convention." *Electrician and Electrical Engineer*, October 1886, 388.

"Detroit Meeting of the National Electric Light Association." *Electrician and Electrical Engineer*, October 1886, 361.

"Edison at the Pan-American Exposition," *Western Electrician* 29:7, August 17, 1901, 103

"Edison's Prophecy: A Duplex, Sleepless, Dinnerless World." *Literary Digest*, November 2, 1923, 966–968.

"Electric Lighting in Boston." *Electrical World and Engineer*, September 19, 1903.

"Electric Lighting in the City." *Engineering* 31 (April 1, 1881): 337.

"Electrical Miracles of Tomorrow." *Literary Digest*, July 18, 1925, 24.

"Electrical News and Notes." *Electrician and Electrical Engineer*, October 1886, 389.

"Electricity in the West." *Frank Leslie's Popular Monthly* 34 (November 1892).

"Historical Sketch of the Foreign Business of the General Electric Company." *General Electric Digest* 2, no. 4 (1922): 5–8; no. 5 (1992): 15–19; no. 6 (1922): 4–9.

"How to Celebrate a Reform Act." *Examiner*, June 24, 1832.

"Lighting the City by Towers." *Hazard's Register of Pennsylvania*, January 25, 1834, 55.

"London Street Lighting." *Engineer* 80 (August 2, 1895): 112.

Narrative of a Tour of Observation Made during the Summer of 1817. Philadelphia: S. A. Mitchell and H. Ames, 1818.

"Notes on Municipal Government: The Relation of the American Municipalities to the Gas and Electric Light Service: A Symposium," *Annals of the American Academy of Political and Social Science* 27 (January 1906): 200–233.

Official Guide of the Ohio Valley and Central States Exposition. Cincinnati: John F. C. Mullen, 1888.

"Outdoor Advertising in France." *Journal of the Society of Arts*, October 4, 1907, 1058.

"Outdoor Advertising in Germany." *Journal of the Society of Arts*, April 10, 1907, 606–607.

"Pageantry for Returning Heroes." *Literary Digest*, April 12, 1919, 26–27.

"Pillar of Fire." *Omaha Daily Bee*, July 13, 1887, 6.

Proceedings of the National Electric Light Association, 1885. Baltimore: Baltimore Publishing Company, 1885.

Proceedings of the National Electric Light Association, 1886. Baltimore: Baltimore Publishing Company, 1886.

Proceedings of the National Electric Light Association, 1903, Appendix B, 22–24.

"Proposed Electric Light Tower at New Orleans." *Scientific American* 46 (March 18, 1882): 11, 159.

"Report on Detroit," *The Electrical World*, May 5, 1888, 233.

"Some Features of Cost Keeping and Accounting at Willow Grove Park, Philadelphia." *Electric Railway Journal* 33, no. 9 (February 27, 1909), 370–371.

"Spectacular Electric Illumination during the Hudson-Fulton Celebration." *Electrical World* 54, no. 6 (1909): 290.

"Street Illuminations of St. Louis." *Frank Leslie's Popular Magazine* 36, no. 3 (September 1893): 25.

"The Cotton Centennial Exhibition at New Orleans." *American Architect and Building News*, April 8, 1885, 171.

The Columbian Gallery: A Portfolio of Photographs from the World's Fair. Chicago: Werner Company, 1894.

"The Electrical Exposition." *Harper's Weekly*, September 13, 1884.

"The Electrical Utility Exhibits at the New York World's Fair." *Edison Electric Institute Bulletin* 7, no. 3 (1939): 81–87.

"The International Electrical Exposition, Philadelphia." *Scientific American*, September 27, 1884, 192.

"The Lights of Broadway. *Literary Digest*, November 18, 1913.

"The Model City at St. Louis." *Charities* 12 (February 6, 1904): 126–127.

"The Presidential Tour." *Harper's Weekly*, October 15, 1887, 143.

"The Veiled Prophet." *Harper's Weekly*, October 16, 1886.

The Wonders of the World's Fair. Buffalo: Barnes, Hengerer, and Co., 1901.

"Tower System for Electric Arc Lighting." *American Architect and Building News*, June 10, 1882, 11, 227.

"Unveiling of Large Electric Flag Crowning Event." *Signs of the Times*, May 1916, 17.

"Veiled Prophet." *Harper's Weekly*, October 16, 1880, 667.

"Wonderland in Electric Building." *Current Literature* 14, no. 1 (September 1893): 21.

Ackley, Laura A. *San Francisco's Jewel City*. Berkeley: California Historical Society, 2015.

Adams, Charles Francis. *An Autobiography*. New York: Chelsea House, 1983.

Adams, Henry. *The Education of Henry Adams*. In *The Works of Henry Adams*. Vol. 1, 715–1192. New York: Library of America, 1982.

Adams, W. E. *Our American Cousins*. London: Walter Scott, 1883.

Adas, Michael. *Machines as the Measure of Men*. Ithaca, NY: Cornell University Press, 1989.

Anderson, O. P. "Brief Outline of Electric Sign History." *Signs of the Times*, May 1916, 18–19.

Archer, William. *America Today*. New York: Scribner's Sons, 1899.

Armengaud, Marc, Matthias Armengaud, and Alessandra Cianchetta. *Nightscapes: Paisajes Nocturnos*. Barcelona: Editorial Gustavo Gili, 2009.

Atless, James. *Nocturne: A Journey in Search of Moonlight*. Chicago: University of Chicago Press, 2011.

Bach, Penny Balkin. "To Light up Philadelphia: Lighting, Public Art, and Public Space." *Art Journal* 48, no. 4 (Winter 1989): 324–330.

Baker, Laura E. "Public Sites versus Public Sights: The Progressive Response to Outdoor Advertising and the Commercialization of Public Space." *American Quarterly* 59, no. 4 (December 2007): 1187–1213.

Baldwin, Peter C. *In the Watches of the Night: Life in the Nocturnal City, 1820–1930*. Chicago: University of Chicago Press, 2012.

Bangs, John Kendrick. "Dewey's Return." *Harper's Weekly* 43, September 23, 1899, 935.

Barnaby, Alice. *Light Touches: Cultural Practices of Illumination, 1800–1900*. London: Routledge, 2017.

Barnes, James. "The Story of Dewey's Welcome Home." *Outlook* (October 1899): 299–312.

Barrett, J. P. *Electricity at the Columbian Exposition*. Chicago: R. R. Donnelley and Sons, 1894.

Barrett, Myrtle I. *Our Wondrous Trip*. Blackwell, OK: Tribune Press, 1914.

Barry, Richard. *The Grandeurs of the Exposition*. Buffalo: Robert Allan Reid Pub., 1901.

Beauchamp, Scott. "The Mystery of St. Louis's Veiled Prophet." *Atlantic*, September 2, 2014.

Beaumont, Matthew. *Night Walking: A Nocturnal History of London*. London: Verso, 2015.

Bedarida, Francois, and Anthony R. Sutcliffe. "The Street in the Structure and Life of the City: Reflections on Nineteenth-Century London and Paris." In *Modern Industrial Cities: History, Policy, and Survival*, edited by Bruce M. Stave, 27–31. Beverly Hills, CA: Sage, 1981.

Bell, Louis. "Elements of Illumination." *Electrical World and Engineer* 29 (August 10, 1901): 221.

Beltran, Alain, and Patrice A. Carré. *La fée et la servante: La societé française face a l'électricité*. Paris: Éditions Belin, 1991.

Binder, Frederick Moore. "Gas Light." *Pennsylvania History* 22, no. 4 (1955): 359–373.

Binder, Frederick Moore, and David M. Reimers. *All the Nations under Heaven: An Ethnic and Racial History of New York City*. New York: Columbia University Press, 1995.

Biringuccio, Vannoccio. *La Pyrotechnie*. Paris: Chez Guillaume Iullian, 1572.

Blumberg, Barbara. "A National Monument Emerges." In *Liberty: The French-American Statue in Art and History*, edited by Pierre Provoyeur and June Ellen Hargrove, 210–217. New York: Harper and Row, 1986.

Bolton, Kate. "The Great Awakening of the Night: Lighting America's Streets." *Landscape* 23 (1979): 41–47.

Bouman, Mark J. "'The Good Lamp Is the Best Police': Metaphor and Ideologies of the Nineteenth-Century Urban Landscape." *American Studies* 32, no. 2 (Fall 1991): 63–78.

Bowers, Brian. *A History of Light and Power*. Stevenage, UK: Peter Peregrinus, 1982.

Boyer, Paul. *Urban Masses and Moral Order in America, 1820–1920*. Cambridge, MA: Harvard University Press, 1978.

Brechin, Grey. "Sailing to Byzantium: The Architecture of the Panama Pacific International Exposition." *California History* 62, no. 2 (1983): 106–121.

Bressani, Martin. 2015. "Paris." In *Cities of Light: Two Centuries of Urban Illumination*, edited by Sandy Isenstadt, Margaret Maile Petty, and Dietrich Neuman, 28–36. London: Routledge.

Bright, Arthur A., Jr. *The Electric Lamp Industry*. New York: Macmillan, 1949.

Brockett, Oscar G. *History of the Theater*. Boston: Allyn and Bacon, 1977.

Brown, Shannon Allen. "Annihilating Time and Space: The Electrification of the United States Army, 1875–1920." PhD diss., University of California at Santa Cruz, June 2000.

Brush, Edward. "Electrical Illumination at the Pan-American Exposition." Supplement, *Scientific American* 51 (January 19, 1901): 20943–20944.

Buck-Morss, Susan. *The Dialectics of Seeing: Walter Benjamin and the Arcades Project*. Cambridge, MA: MIT Press, 1991.

Burne-Jones, Philip. *Dollars and Democracy*. New York: D. Appleton Company, 1904.

Burnham, Daniel. *Report on a Plan for San Francisco*. San Francisco: City of San Francisco, 1905.

Burnham, Daniel, and Edward A. Bennett. *Plan of Chicago*. Chicago: Commercial Club, 1909.

Burrows, Edwin G., and Mike Wallace. *Gotham: A History of New York City to 1898*. New York: Oxford University Press, 1999.

Byatt, I.C.R. *The British Electrical Industry, 1875–1914*. Oxford: Clarendon Press, 1979.

Carletta, David Mark. "The Triumph of American Spectacle: New York City's 1892 Columbian Celebration." *Material Culture* 40, no. 1 (Spring 2008): 19–40.

Carlson, W. Bernard. *Tesla: Inventor of the Electrical Age*. Princeton, NJ: Princeton University Press, 2013.

Cather, Willa. "Behind the Singer Tower." In *The Collected Short Fiction of Willa Cather*, 43–54. Lincoln: University of Nebraska Press, 1965.

Cavling, Henrik. *Fra Amerika*. Vol. 1. Copenhagen: Gyldendalske Boghandels Forlag, 1897.

Chesterton, G. K. *What I Saw in America*. London: Hodden and Stoughton, 1922.

Cleveland, Cutler J. *Concise Encyclopedia of Energy History*. Cambridge, MA: Academic Press, 2009.

Committee of Board of Legislation of Cincinnati. "Report on Street Lighting in Various Large Cities in the United States." *Proceedings of the National Electric Light Association* 24 (1901): 423–451.

Coopersmith, Jonathan. *The Electrification of Russia, 1880–1926*. Ithaca: Cornell University Press, 1992.

Coopman, David T. *The History of Rock Island County*. Mt. Pleasant, SC: Arcadia Publishing, 2008.

Cox, James. *St. Louis through a Camera*. Saint Louis: Woodward and Tiernan Printing Co., 1892.

Creel, George. *How We Advertised America.* New York: Harper and Brothers, 1920.

Davis, Richard Harding. *The Great Streets of the World.* New York: Charles Scribner's, 1892.

Dewey, Stoddard. "Street Lamps of Paris," *Youth's Companion*, August 15, 1895, 387.

Dickens, Charles. *The Uncommercial Traveler.* New York: Charles Scribner's Sons, 1905.

Dickerson, A. F. "Spectacular Lighting." *Proceedings of the National Electric Light Association Convention* 47 (1924): 484–486.

Dodderer-Winkler, M. *Magnificent Entertainment: Temporary Architecture for Georgian Festivals.* New Haven, CT: Yale University Press, 2013.

Domosh, Mona. *Invented Cities: The Creation of Landscape in Nineteenth-Century New York and Boston.* New Haven, CT: Yale University Press, 1996.

Dow, Alex. "Public Lighting in Relation to Public Ownership and Operation." In *Proceedings of the National Electric Light Association, 21st Convention, 1898*, 93–116. New York: James Kepster Printing, 1898.

Dreiser, Theodore. *Newspaper Days.* Philadelphia: University of Pennsylvania Press, 1991.

Ekirch, A. Roger. *At Day's Close: A History of Nighttime.* London: Phoenix Books, 2006.

Eshelman, C. L. "Modern Streetlighting." *American City* 6 (1912): 510–517.

Everett, Marshall. *The Book of the Fair, the Greatest Exposition the World Has Ever Seen: A Panorama of the St. Louis Exposition.* Philadelphia: Henry Neil, 1904.

Fenske, Gail. *The Skyscraper and the City: The Woolworth Building and the Making of Modern New York.* Chicago: Chicago University Press, 2008.

Fishback, Frederick L. "Chronicler's Report for 1915." *Records of the Columbia Historical Society, Washington, D.C.* 19 (1916): 222–223.

Fisher, Philip. *Still the New World.* Cambridge, MA: Harvard University Press, 1999.

Francis, David R. *The Universal Exposition of 1904.* Saint Louis: Louisiana Purchase Exposition Company, 1913.

Freeberg, Ernest. *The Age of Edison: Electric Light and the Invention of Modern America.* New York: Penguin, 2013.

Fri, Robert W. "The Alternative Energy Future: The Scope of Transition." *Daedalus* 142, no. 1 (Winter 2013): 5–7.

Friedel, Robert, and Paul Israel. *Edison's Electric Light: Biography of an Invention.* New Brunswick, NJ: Rutgers University Press, 1985.

Gibson, Jane Mark. "The International Electrical Exhibition of 1884 and the National Conference of Electricians." MA thesis, University of Pennsylvania, 1984.

Gilbert, James. *Perfect Cities: Chicago's Utopias of 1893.* Chicago: University of Chicago Press, 1991.

Gilbert, James. *Whose Fair? Experience, Memory, and the History of the Great St. Louis Exposition.* Chicago: University of Chicago Press, 2009.

Gilchrist, J. F. "Electric Signs." *Proceedings of the National Electric Light Association* 29, no. 1 (1906): 318–342.

Glassberg, David. "Public Ritual and Cultural Hierarchy: Philadelphia's Civic Celebrations at the Turn of the Twentieth Century." *Pennsylvania Magazine of History* 107 (1983): 421–448.

Goethe, Johann Wolfgang von. *Italian Journey.* Trans. W. H. Auden and Elizabeth Meyer. San Francisco: North Point Press, 1992.

Goulet, Nicolas. *Fêtes à l'occasion du mariage des S. M. Napoléon, empereur des Français, roi d'Italie, avec Marie-Louise, archiduchesse d'Autriche: Recueil de gravures au trait, représentant les principales décorations d'architecture et de peinture, et les illuminations les plus remarquables auxquelles ce mariage a donné lieu.* Paris: Chez L.Ch. Soyer, 1810.

Grant, Robert. "Notes on the Pan-American Exposition." *Cosmopolitan* 31, no. 5 (September 1901): 450–464.

Habermas, Jürgen. "The Public Sphere." *New German Critique* 5 (1974): 49–55.

Hall, Edward H. *The Hudson Fulton Celebration.* Albany: State of New York, 1910.

Hammond, John Winthrop. *Men and Volts.* Philadelphia: Lipppincott, 1941.

Hannah, Leslie. *Electricity before Nationalisation: A Study of the Development of the Electricity Supply Industry in Britain to 1918.* Baltimore: Johns Hopkins University Press, 1979.

Harrison, Wallace. "The Importance of Artificial Light to Architecture." *Transactions of the Illuminating Engineering Society* (May 1930): 475–478.

Haskell, Douglas. "Architecture, the Bright Lights." *Nation* 132 (January 14, 1931): 55–56.

Haskin, Frederic J. *The Panama Canal.* New York: Doubleday, Page, and Co., 1913.

Hatton, Joseph. *Henry Irving's Impressions of America, Narrated in a Series of Sketches, Chronicles, and Conversations.* Boston: J. R. Osgood and Company, 1884.

Hausman, William J., Mira Wilkins, and John L. Neufeld, "Global Electrification: Multinational Enterprise in the History of Light and Power, 1880s–1814." *Revue économique* 58, no. 1 (January 2007): 175–190.

Hawthorne, Nathaniel. "The Scarlet Letter." In *Complete Novels of Nathaniel Hawthorne,* edited by Norman Holmes Pearson, 115–345. New York: Library of America, 1983.

Haynes, James B., ed. *History of the Trans-Mississippi and International Exposition of 1898*. Omaha: privately printed, 1910.

Haywood, William. (Report to the Streets Committee). *Results of the Electric Lighting of Public Ways within the City of London in 1881–1882*. London: Charles, Skipper, and East, 1882.

Heap, Chad. *Sexual and Racial Encounters in American Nightlife, 1885–1940*. Chicago: University of Chicago Press, 2009.

Heinze, Andrew. *Adapting to Abundance: Jewish Immigrants, Mass Consumption, and the Search for American Identity*. New York: Columbia University Press, 1990.

Helm, Thomas B. *History of Wabash County*. Chicago: John Morris, 1884.

Henderson, Mary C. *Theater in America*. New York: Harry N. Abrams, 1986.

Homann, Joachim, ed. *Night Vision: Nocturnes in American Art, 1860–1960*. Munich: Delmonico Books, 2014.

Huebinger, Melchior. *First Album of the City of Davenport, Iowa*. Davenport: Huebinger's Photographic Gallery, 1887.

Hughes, Thomas P. "British Electrical Industry Lag: 1882–1888." *Technology and Culture* 3, no. 1 (1962): 27–44.

Hughes, Thomas P. *Networks of Power*. Baltimore: Johns Hopkins University Press, 1983.

Huhtamo, Erkki. "The Sky Is Not the Limit: Envisioning the Ultimate Public Media Display." *Journal of Visual Culture* 8 (2009): 329.

Hunter, Louis C. *Steam Power*. Vol. 2, *A History of Industrial Power in the United States, 1780–1930*. Charlottesville: University Press of Virginia, 1985.

Inwood, Stephen. *City of Cities: The Birth of Modern London*. London: Macmillan, 2005.

Isenstadt, Sandy. "Los Angeles." In *Cities of Light: Two Centuries of Urban Illumination*, edited by Sandy Isenstadt, Margaret Maile Petty, and Dietrich Neumann, 51–57. London: Routledge, 2015.

Isenstadt, Sandy, Margaret Maile Petty, and Dietrich Neumann, eds. *Cities of Light: Two Centuries of Urban Illumination*. London: Routledge, 2015.

Israel, Paul. *Edison: A Life of Invention*. New York: John Wiley, 1998.

Jacobs, Jane. "Downtown Is for People." In *The Exploding Metropolis*, edited by William H. Whyte, 157–184. Berkeley: University of California Press, 1993.

Jacobs, Jane. *Life and Death of American Cities*. New York: Vintage Books, 1992.

Jakle, John A. *City Lights: Illuminating the American Night*. Baltimore: Johns Hopkins University Press, 2001.

Johnson, Hildegard Binder. "Towards a National Landscape." In *The Making of the American Landscape*, edited by Michael Conzon, 127–145. New York: HarperCollins, 1994.

Jones, Christopher. *Routes of Power: Energy and Modern America.* Cambridge, MA: Harvard University Press, 2014.

Jones, Franklin D. *Engineering Encyclopedia.* New York: Industrial Press, 1941.

Josephson, Matthew. *Edison.* New York: McGraw Hill, 1959.

Kaempffert, Walter. *Ornamental Street-Lighting: A Municipal Investment and Its Return.* Cleveland: National Electric Light Association, 1912.

Kahn, Judd. *Imperial San Francisco.* Lincoln: University of Nebraska Press, 1979.

Kasson, John. *Amusing the Million.* New York: Hill and Wang, 1978.

Keating, Paul W. *Lamps for a Brighter America.* New York: McGraw Hill, 1954.

Kendall, John. *History of New Orleans.* Chicago: Lewis Publishing Company, 1922.

Kenny, D. J. *Illustrated Cincinnati.* Cincinnati: Robert Clarke and Co., 1875.

King, Moses. *King's Handbook of New York.* Boston, 1892.

Kline, Ronald. *Steinmetz: Engineer and Socialist.* Baltimore: Johns Hopkins University Press, 1992.

Klingenberg, G. "Electricity Supply of Large Cities." *Journal of the Institution of Electrical Engineers* 52 (1914): 123–149.

Koester, Frank. "Electric Lighting, Police, and Fire Alarms." *American City*, February 1913, 33–38.

Koron, Joseph J., Jr. *The American Skyscraper: A Celebration of Height.* Wellesley, MA: Brandon Books, 2008.

Koslofsky, Craig. *Evening's Empire: A History of Night in Early Modern Europe.* Cambridge: Cambridge University Press, 2011.

Lacombe, C. F. "Street Lighting Systems and Fixtures in New York City." *American City*, May 1913, 516–519.

Laird, Pamela Walker. *Advertising Progress: American Business and the Rise of Consumer Marketing.* Baltimore: Johns Hopkins University Press, 1998.

Larwood, Jacob. *The History of Signboards, from the Earliest Times to the Present Day.* London: John Camden Hotten, 1864.

Layton, Edwin. *The Revolt of the Engineers.* Cleveland: Case Western Reserve University Press, 1971.

Leach, William. *Land of Desire: Merchants, Power, and the Rise of a New American Culture.* New York: W. W. Norton, 1989.

Lears, T. J. Jackson. *No Place of Grace.* New York: Pantheon Books, 1981.

Levinson, J. C., Ernest Samuels, Charles Vandersee, and Viola H. Winner, eds. *The Letters of Henry Adams, Vol. 4: 1892–1918*. Cambridge, MA: Harvard University Press, 1988.

Lippincott, Wimot. *Outdoor Advertising*. New York: McGraw Hill, 1923.

Luckiesh, Matthew. *Artificial Light: Its Influence on Civilization*. New York: Century Co., 1920.

Lynn, Michael R. "Sparks for Sale: The Culture and Commerce of Fireworks in Early Modern France." *Eighteenth Century Life* 30, no. 2 (2006): 74–97.

Mandell, Richard D. *Paris, 1900*. Toronto: University of Toronto Press, 1967.

Marer, Lise-Lone. *David Belasco: Naturalism in the American Theater*. Princeton, NJ: Princeton University Press, 1975.

Martin, T. Commerford, and Luther Stieringer. "On the Electric Lighting of the World's Fair." In *Proceedings of the National Electric Light Association*, 189–195. New York: James Hempster Printing Co., 1894.

Marvin, Carolyn. *When Old Technologies Were New*. New York: Oxford University Press, 1988.

Massey, Jonathan. "Organic Architecture and Direct Democracy: Claude Bragdon's Festivals of Song and Light." *Journal of the Society of Architectural Historians* 65, no. 4 (December 2006): 578–613.

McAllister, E. J. "Possibilities of Sign and Decorative Lighting." *Proceedings of the National Electric Light Association* 25 (1902): 320–329.

McCabe, James D. *New York by Gaslight*. New York: Arlington House, 1984. First published 1882.

McLaren, John, and W. D'Arcy Ryan. "Coloring a City." In *The Story of the Exposition*, Vol. 1, edited by Frank Morton Todd, 347–353. New York: G. P. Putnam's, 1921.

McNamara, Brooks. *Day of Jubilee: The Great Age of Public Celebrations in New York, 1788–1909*. New Brunswick, NJ: Rutgers University Press, 1997.

McShane, Clay. *Down the Asphalt Path*. New York: Columbia University Press, 1994.

Meier, Josiane, Ute Hasenöhrl, Katharina Krause, and Merle Pottharst, eds. *Urban Lighting, Light Pollution, and Society*. New York: Routledge, 2015.

Melosi, Martin V. "Energy Transitions in the Nineteenth-Century Economy." In *Energy and Transport*, edited by G. H. Daniels and Mark Rose, 55–69. Thousand Oaks, CA: Sage Publications, 1982.

Merrick, Samuel Vaughan. *Report of the Committee to Whom Was Referred Sundry Memorials against Lighting the City with Gas*. Philadelphia: Lydia R. Bailey, 1833.

Milder, Robert. *Reimagining Thoreau.* Cambridge: Cambridge University Press, 1995.

Millar, Preston S. "Wartime Lighting Economy." In *Proceedings of the Second War Convention, National Electric Light Association, Atlantic City, July 13–14,* 331–335. New York: National Electric Light Association, 1918.

Mills, E. A. "The Development of Electric Sign Lighting." *Transactions of the Illuminating Engineering Society* 15, no. 6 (August 30, 1920): 363–371.

Missal, Alexander. *Seaway to the Future: American Social Visions and the Construction of the Panama Canal.* Madison: University of Wisconsin Press, 2008.

Mohr, Nicolaus. *Excursion through America.* Chicago: Lakeside Press, 1973. Translation of original, Berlin, 1884.

Moncel, Comte Th. Du, and William Preece. *Incandescent Lights, with Particular Reference to the Edison Lamps at the Paris Exhibition.* New York: Van Nostrand, 1882.

Moore, Mark, and Karl Strand. "Preservation Study of the Moonlight Towers, Austin, Texas." *APT Bulletin: The Journal of Preservation Technology* 23, no. 1 (1991): 29–38.

Moore, Sarah J. *Empire on Display: San Francisco's Panama Pacific International Exposition.* Norman: University of Oklahoma Press, 2013.

Mortimer, G. W. *Pyrotechny, or a Familiar System of Recreative Fireworks.* London: James S. Hodson, 1824.

Nardis, Fred. *Wonder Shows.* New Brunswick, NJ: Rutgers University Press, 2005.

Nasaw, David. *Going Out: The Rise and Fall of Public Amusements.* New York: Basic Books, 1993.

Nathan-Garner, Laura. *Insider's Guide to Houston.* Guilford, CT: Globe-Pequot Press, 2012.

Nead, Lynda. *Victorian Babylon.* New Haven: Yale University Press, 2000.

Neumann, Dietrich. *Architecture of the Night.* Munich: Prestel, 2002.

New York Edison Company. *Forty Years of Edison Service, 1882–1922.* New York: New York Edison Company, 1922.

Nye, David E. *America as Second Creation.* Cambridge, MA: MIT Press, 2003.

Nye, David E. *American Technological Sublime.* Cambridge, MA: MIT Press, 1994.

Nye, David E. *America's Assembly Line.* Cambridge, MA: MIT Press, 2013.

Nye, David E. *Consuming Power: A Social History of American Energies.* Cambridge, MA: MIT Press, 1998.

Nye, David E. *Electrifying America: Social Meanings of New Technology.* Cambridge, MA: MIT Press, 1990.

Nye, David E. "Implementing a New Energy Regime in Housing." In *Energy Accounts: Architectural Representations of Energy, Climate, and the Future*, edited by Dan Willis, William W. Braham, Katsuhiko Muramoto, and Daniel A. Barber, 32–40. London: Routledge, 2017.

Nye, David E. *Technology Matters: Questions to Live With*. Cambridge: MIT Press, 2006.

Nye, David E. *When the Lights Went Out*. Cambridge, MA: MIT Press, 2010.

O'Dea, William. *The Social History of Lighting*. New York: Macmillan, 1958.

Onuf, Peter S. "Liberty, Development, and Union: Visions of the West in the 1780s." *William and Mary Quarterly* 43, no. 2 (1986): 179–233.

Otter, Christopher. *The Victorian Eye: A Political History of Light and Vision in Britain, 1800–1910*. Chicago: Chicago University Press, 2008.

Otter, Christopher. "Cleansing and Clarifying: Technology and Perception in Nineteenth-Century London." *Journal of British Studies* 43, no. 1 (January 2004): 40–64.

Palmer, Ray. "Municipal Lighting Rates." *Annals of the American Academy*, January 1915.

Parsons, R. H. *The Early Days of the Power Station Industry*. Cambridge: Cambridge University Press, 1940.

Passer, Harold C. *The Electrical Manufacturers*. Cambridge, MA: Harvard University Press, 1953.

Penzel, Frederick. *Theatre Lighting before Electricity*. Middletown, CT: Wesleyan University Press, 1978.

Peterson, Jon A. *The Birth of City Planning in the United States, 1840–1917*. Baltimore: Johns Hopkins University Press, 2003.

Pincus, Henry. "Common Errors in Park Construction." *Street Railway Journal*, May 5, 1900, 461.

Plotnick, Rachel. "At the Interface: The Case of the Electric Push Button, 1880–1923." *Technology and Culture* 53, no. 4 (October 2012): 815–845.

Porter, Roy. *London: A Social History*. Cambridge, MA: Harvard University Press, 1994.

Pound, Ezra. "Patria Mia." In *Selected Prose, 1909–1965*, 99–141. New York: New Directions, 1973.

Pratt, Harold. *The Electric City: Energy and the Growth of the Chicago Area*. Chicago: University of Chicago Press, 1991.

Preece, W. H. "Our Lights as Others See Them." *Electrical World* (December 27, 1884): 264.

Preece, W. H. "Public Lighting in America," *Journal of the Society of Arts* (December 5, 1884): 66–73.

Presbury, Frank. *The History and Development of Advertising.* Garden City, NY: Doubleday, Doran, and Co., 1929.

Rae, Frank B., and George Williams. "Creating Demands for Electricity." In *Proceedings of the 31st Convention of the National Electric Light Association, Chicago*, 751–760. New York: National Electric Light Association, 1908.

Reps, John W. *The Making of Urban America: A History of City Planning in the United States.* Princeton, NJ: Princeton University Press, 1992.

Riis, Jacob. *How the Other Half Lives.* New York: Charles Scribner's, 1890.

Robinson, Charles Mulford. "Improvement in City Life: Aesthetic Progress." *Atlantic Monthly* 83 (June 1899): 771–785.

Robinson, Charles Mulford. *The Improvement of Towns and Cities; or, the Practical Basis of Civic Aesthetics.* New York: G. P. Putnam's, 1901.

Roemer, Kenneth M. *The Obsolete Necessity: America in Utopian Writings, 1888–1900.* Kent, OH: Kent State University Press, 1981.

Rose, Mark. *Cities of Heat and Light: Domesticating Gas and Electricity in Urban America.* University Park, PA: Pennsylvania State University Press, 1995.

Rossell, Edward Graham Daves. "Compelling Vision: From Electric Light to Illuminating Engineering, 1880–1940." PhD diss., University of California at Berkeley, 1998.

Ruggieri, Claude-Fortuné. *Elemens de pyrotechnie.* Paris: Chez Barba, 1811.

Ryan, W. D'Arcy. "The Illumination of the Panama-Pacific Exposition." *General Electric Review* 18 (1915): 579–586.

Rydell, Robert. *All the World's a Fair.* Chicago: University of Chicago Press, 1984.

Sandweiss, Eric. *St Louis: The Evolution of an American Urban Landscape.* Philadelphia: Temple University Press, 2001.

Schivelbusch, Wolfgang. *Disenchanted Night: The Industrialization of Light in the Nineteenth Century.* Berkeley: University of California Press, 1988.

Schlereth, Thomas J. "Columbia, Columbus, and Columbianism." *Journal of American History* 79, no. 3 (December 1992): 937–968.

Schlereth, Thomas J. *Victorian America.* New York: HarperCollins, 1991.

Schlor, Joachim. *Nights in the Big City.* London: Reaktion Books, 1998.

Schott, Dieter. 2008. "Empowering European Cities: Gas and Electricity in the Urban Environment." In *Urban Machinery: Inside Modern European Cities*, edited by Mikael Hård and Thomas J. Misa, 165–186. Cambridge, MA: MIT Press.

Scranton, Philip. *Endless Novelty: Specialty Production and American Industrialization, 1865–1925*. Princeton, NJ: Princeton University Press, 1997.

Sears, John. *Sacred Places: American Tourist Attractions in the Nineteenth Century*. New York: Oxford University Press, 1989.

Seelye, John. "'Rational Exultation': The Erie Canal Celebration." *Proceedings of the American Antiquarian Society* 94, no. 2 (1985): 241–267.

Segal, Howard P. 1994. "Edward Bellamy and Technology." In *Future Imperfect: The Mixed Blessings of Technology in America*, 101–116. Amherst: University of Massachusetts Press, 1994.

Segal, Howard P. 1986. "The Technological Utopians." In *Imagining Tomorrow: History, Technology, and the American Future*, edited by Joseph J. Corn, 119–136. Cambridge, MA: MIT Press, 1988.

Sharpe, C. Melvin. "Brief Outline of the History of Electric Illumination in the District of Columbia." *Records of the Columbia Historical Society* 48–49 (1946–1947): 191–207.

Sharpe, William Chapman. *New York Nocturne: The City after Dark in Literature, Painting, and Photography*. Princeton, NJ: Princeton University Press, 2008.

Shiman, Daniel R. "Explaining the Collapse of the British Electricity Supply Industry in the 1880s: Gas vs. Electricity Prices." *Business and Economic History* 22 (1993): 318–327.

Shultz, Quentin J. "Legislating Morality: The Progressive Response to American Outdoor Advertising, 1900–1917." *Journal of Popular Culture* 17, no. 4 (Spring 1984): 37–44.

Simpson, Thomas Bartlett. *Gas-Works: The Evils Inseparable from Their Existence in Populous Places, and the Necessity of Removing Them from the Metropolis*. London: William Freedman, 1866.

Smith, Matthew Hale. *Sunshine and Shadow in New York*. Hartford: J. B. Burr, 1869.

Spaulding, Russell. "Display Lighting, Signs, and Decorative Light." *Proceedings of the National Electric Light Association* 25 (1902): 311–317.

Spencer, Thomas. *The St. Louis Veiled Prophet Celebration: Power on Parade, 1877–1995*. Columbia: University of Missouri Press, 2000.

Starr, Tama, and Edward Hayman. *Signs and Wonders: The Spectacular Marketing of America*. New York: Doubleday Books, 1998.

Stieringer, Luther. "The Evolution of Exposition Lighting." *Western Electrician* 29, no. 12 (September 21, 1901): 187–190.

Stevenson, Robert Louis. "A Plea for Gas Lamps." In *Virginibus Puerisque*, 249–256. Boston: Small, Maynard, and Co., 1907 [originally published in *The London Magazine*, April 27, 1878].

Stradling, David, and Peter Thorsheim. "The Smoke of Great Cities: British and American Efforts to Control Air Pollution." *Environmental History* 4, no. 1 (January 1999): 6–31.

Talbot, Frederick Arthur Ambrose. *Electrical Wonders of the World*. Vol. 1. Toronto: Cassell, 1921.

Tarr, Joel, and Gabriel Dupuy. *Technology and the Rise of the Networked City in Europe and America*. Philadelphia: Temple University Press, 1988.

Tauranac, John. *The Empire State Building*. New York: Scribner's, 1995.

Thoreau, Henry David. "Night and Moonlight." *Atlantic Monthly Magazine*, November 1863, 579–583.

Tichi, Cecelia. *Shifting Gears: Technology, Literature, Culture in Modernist America*. Chapel Hill: University of North Carolina Press, 1987.

Tocco, Peter. "The Night They Turned the Lights on in Wabash." *Indiana Magazine of History* 95, no. 4 (1999): 350–363.

Todd, Frank Morton. *The Story of the Exposition*. Vol. 2. New York: Putnam's Sons, 1921.

Tomory, Leslie. "The Environmental History of the Early British Gas Industry, 1812–1830." *Environmental History* 17, no. 1 (January 2012): 29–54.

Tomory, Leslie. *Progressive Enlightenment: The Origins of the Gaslight Industry, 1780–1820*. Cambridge, MA: MIT Press, 2012.

Traub, James. *The Devil's Playground: A Century of Pleasure and Profit in Times Square*. New York: Random House, 2004.

Trouillot, Michel-Rolph. "Good Day Columbus: Silences, Power, and Public History (1492–1892)." *Public Culture* 3, no. 1 (Fall 1990): 1–24.

Turner, C. Y. "The Color Scheme." In *Art Handbook: Official Catalogue of Architecture and Sculpture and Art, Pan-American Exposition*. Buffalo: David Gray, 1901. http://panam1901.org/documents/original_sources/art_handbook_gray.pdf.

Twain, Mark. *A Tramp Abroad*. New York: Harper and Brothers, 1907.

Twain, Mark. *Following the Equator*. New York: Dover, 1989.

Twain, Mark. *Historical Romances: The Prince and the Pauper*. New York: Library of America, 1994.

Twain, Mark. *Life on the Mississippi*. Boston: James R. Osgood, 1883.

Twain, Mark. *Mark Twain's Notebooks and Journals*, Vol. III, edited by Robert Park Browning, Michael B. Frank, and Lin Salama. Berkeley: University of California Press, 1979.

Twain, Mark. *The Innocents Abroad*. New York: Signet, 1980.

Urbanitzky, Alfred Ritter von. *Electricity in the Service of Man*. London: Casells and Co., 1886.

Van Schaack, Eric. "The Division of Pictorial Publicity in World War I." *Design Issues* 22, no. 1 (Winter 2006): 32–45.

Vegas, Fernando, and Camilla Mileto. "World's Fairs: Language, Interpretation, and Display." *Change over Time* 3, no. 2 (2013): 174–187.

Vredenburgh, LaRue. "Sign and Decorative Lighting." *Proceedings of the National Electric Light Association* 26 (1903): 344–350.

Wainwright, Nicholas B. *History of the Philadelphia Electric Company, 1881–1961*. Philadelphia: Philadelphia Electric Company, 1961.

Walker, John Brisben. "The City of the Future—A Prophecy," *Cosmopolitan* 31, no. 3 (1901): 473–475.

Wallace, Alfred Russel. *The Progress of the Country*. New York: Harper Brothers, 1901.

Wells, H. G. *The Future in America*. Leipzig: Bernhard Tauchnitz, 1907.

Werrett, Simon. *Fireworks: Pyrotechnic Arts and Sciences in European History*. Chicago: University of Chicago Press, 2010.

Whipple, F. H. *Municipal Lighting*. Detroit: Free Press, 1888.

Whitman, Walt. *Complete Poetry and Prose, Specimen Days*. New York: Library of America, 1982.

Williams, Arthur. "Decorative and Sign Lighting." In *Proceedings of the National Electric Light Association* 26 (1903): appendix A, 2–56.

Wilson, William H. *The City Beautiful Movement*. Baltimore: Johns Hopkins University Press, 1989.

Wrege, Charles D. "J. W. Starr, Cincinnati's Forgotten Genius." *Cincinnati Historical Society Bulletin* 34 (1976): 102–120.

Wrightington, E. N. "Street Lighting with Gas in Europe." *Transactions of the Illuminating Engineering Society* (1908): 533–537.

Ziegler, O. D. "The Living Electric Sign." *Illuminating Engineer*, April 1910, 74–75.

Index